On lampreys and fishes

Developments in environmental biology of fishes 8

Series Editor
EUGENE K. BALON

On lampreys and fishes:
a memorial anthology in honor of Vadim D. Vladykov

Guest editors:

DON E. McALLISTER & EDWARD KOTT

Reprinted from *Environmental biology of fishes* 23 (1–2), 1988
with addition of species and subject index

KLUWER ACADEMIC PUBLISHERS
DORDRECHT / BOSTON / LONDON

Library of Congress Cataloging in Publication Data

On lampreys and fishes : a memorial anthology in honor of Vadim D.
Vladykov / guest editors, Don E. McAllister & Edward Kott.
p. cm. -- (Developments in environmental biology of fishes ;
8)
ISBN 9061936616
1. Lampreys. 2. Fishes. 3. Vladykov, Vadim D. (Vadim Dmitrij),
1898- . I. Vladykov, Vadim D. (Vadim Dmitrij), 1898- .
II. McAllister, D. E. III. Kott, Edward. IV. Series.
QL638.2.O5 1988
597'.2--dc19 88-13271
CIP

ISBN 90-6193-661-6

Published by Kluwer Academic Publishers,
P.O. Box 17, 3300 AA Dordrecht, The Netherlands.

Kluwer Academic Publishers incorporates
the publishing programmes of
Martinus Nijhoff, Dr W. Junk, D. Reidel and MTP Press.

Sold and distributed in the U.S.A. and Canada
by Kluwer Academic Publishers,
101 Philip Drive, Norwell, MA 02061, U.S.A.

In all other countries, sold and distributed
by Kluwer Academic Publishers Group,
P.O. Box 322, 3300 AH Dordrecht, The Netherlands.

Printed in the Netherlands

Contents

Vadim D. Vladykov giving the opening address at the American Society of Ichthyologists and Herpetologists Meeting, Ottawa, 1974.

Environmental Biology of Fishes Vol. 23, No. 1–2, pp. 7–8, 1988

Prelude

A generation apart

Vadim Vladykov became a biology student at the Charles University in Prague, Czechoslovakia, 28 years before me. We both graduated there with similar degrees but three political regimes apart. He arrived in Canada a month after I was born. To complete the reasons for writing this, Vadim had to leave the Prague establishment upon graduation and go as far east in the country as possible (Uzghorod) and so did I (Bratislava); the shrinking distance was determined solely by the new political boundaries. We both ended as Canadians but over 40 years apart, escapees from the same oppressive regime. He was a devoted believer in creation and a member of the Russian Orthodox Church, as I am in evolution and of the pantheist's domain. So we both cherished freedom and disliked cliques, . . . , with similar consequences.

I remember our first meeting on Canadian soil during a wine and cheese party in the National Museum of Natural Sciences, Ottawa. His first words were: 'Why did you come here?! You would have been a leading ichthyologist over there, here you will end up one of many . . .'. Words I failed to understand then, taken by surprise, and that even offended me in my new enthusiasm. But he had known about my arrival some months earlier and his words might have been a result of much thought given to our allochronic but parallel fates. Later, upon more reflection, I thought that he was talking more about himself than me.

When the final reasons for my Canadian destination arrived in the old country (1968) an issue of the official Věstník Československé Společnosti Zoologické (Acta societatis zoologicae Bohemoslovacae 32: 305–393) appeared with articles on fishes by 13 authors and the same line under each author's name: 'Dedicated to the 70th birthday of Professor V.D. Vladykov'. Nothing more, so the censors probably believed they are honoring a Soviet colleague! More reasons for us now to set the record straight.

Vadim is recognized by his peers in the old country as the founder of modern ichthyology there, and his papers were and are referred to and cited constantly. His French and Canadian careers were viewed with respect and even glorified, but his odyssey never understood. When he finally revisited Czechoslovakia with his wife Nina – in 1967 if I remember correctly – we all scrambled to be of assistance.

His main interest then was the huchen or Danubian salmon, *Hucho hucho,* and he wanted to acquire skeletons for osteological studies. The fish, although successfully reared in several hatcheries, was a strictly monitored and highly prized trophy of elite anglers. In over 10 years I managed to preserve a few for much needed mensural and meristic revision at the price of never tasting this renowned delicacy. Vadim not only was given large individuals from a precious parent stock but invited my colleagues to feast on the flesh taken off the skeletons. I had to attend a meeting and again missed the meal.

Even in our new homeland we succeeded more often to miss each other than to meet, excuses notwithstanding. This fact finally mounted into a permanent guilt feeling on my part and led to an insistence on seeing him when in Ottawa early January 1986 (see Epilogue).

A few of us appreciate the hard business of lost language, educational continuity and student brotherhood leading to peer cliques. Vadim had succeeded in becoming a respected scholar in his new homeland in spite of these odds. He must have known that to become 'one of many' feels better among bright peers than among a few barely glowing dispirited survivors. To forfeit work dictated solely by political incompetents in return for freedom of inquiry is not such a bad exchange. But he could not help feeling not entirely right about the advantages others had who were not forced to exchange their 'mental currency'. (After all, economics was the singular North American entity Vadim never managed to cope with well.) It all reminds one of the fate of another 'stranger on the square' – Arthur Koestler. Fate and mentality, for the words Stephen Toulmin (in The Return to Cosmology, 1982, p. 179) felt compelled to write about the latter fits well the former.

> 'The Public Mind has to be rescued from this Dragon; and St. Arthur will always be there, just in time, to cut loose the bonds and carry the heroine off to safety. Still, here again, it is never entirely clear who exactly composes this cabal, or how they supposedly succeed in imposing their authority on those fresh generations of young, hard-nosed, and highly counter-suggestible apprentices who enter the natural sciences every year'.

Vadim, like Arthur, was not easy to understand, but both left a legacy more sound than many who felt superior next to their accented verbal delivery. There is an element of sadness in all this, but also a lesson to be learned. As the Canadian poet Pier Giorgio Di Cicco (1986) said in 'Virgin Science':

... Of course the fish are elusive.
They're not dumb, though we ascribe talents to them
in terms of extra bio-systemic dimensions. It fools
no one. Sleep is similarly bombastic. We pretend to
be scared of the ogres and precipices, and this is
a way of learning how to wake in the new land.
All in all, it was better with you.

Vadim Vladykov would be 90 had he lived to see the publication of this anthology. It was conceived at his bedside 7 days before his passing away and I hope he understood when we told him then our intention to prepare such a volume. He deserved that last good news.

Guelph, January 1988

Eugene K. Balon

Environmental Biology of Fishes Vol. 23, No. 1–2, pp. 9–20, 1988

Vadim Dimitrievitch Vladykov: life of an ichthyologist

Don E. McAllister
Ichthyology Section, National Museum of Natural Sciences, P.O. Box 3443, Station D, Ottawa, Ontario, Canada K1P 6P4

Following a full and productive life Professor Vadim Dimitrievitch Vladykov, internationally known ichthyologist, died in hospital on the night of 14 January 1986 after having lived in his retirement apartment up until his last day. He was 87 years old. Surviving are his two daughters, Nadya Clowes and Kyra Fisher, and four grandchildren. His beloved wife, Nina Formagie, who he married in 1938, predeceased him on the 19th of December 1976.

Dr. Vladykov was born near Kharkov, Ukraine, March 18, 1898 (Gregorian calendar), son of Dimitriy Iv. Vladykov (1863–1955), an archpriest, theologian and missionary of the Russian Orthodox Church, and Nadezhda. Vadim graduated from the Lycée of Akhtyrka in 1917. He was a student at the University of Kharkov from 1917 to 1918 under Prof. P. Sushkin. He enlisted as a private and fought in the White Army under Generals Wrangel and Denikin, caught typhus and while convalescing in the Crimea, studied the fishes of Kertch Channel (= Kerchenskiy Proliv, between the Azov and Black seas). Following the evacuation of the White Army he was discharged with the rank of sublieutenant and surveyed the fishes and herptiles of the Bosphorus in Turkey. The resulting collections were deposited in the French College of Saint-Georges of Constantinople (now Istanbul) and in the Natural History Museum of Prague.

In 1921 he became a biology student at Charles University in Prague, Czechoslovakia and graduated with a RNDr degree (Rerum Naturalium Doctor) in 1925, with a major in zoology and a minor in anthropology. His thesis, published in 1926, dealt with the lateral line system of *Salmo trutta*. It was there that he began his studies on the fishes of Subcarpathian Russia (then in Czechoslovakia, now part of Ukrainian SSR). This led to his appointment in 1927 as director of the Ethnographic Museum of Subcarpathian Russia in Uzhgorod. In the same year he became President of the Society of Aquarists of Subcarpathia. His Subcarpathian studies were published in 1931 and included several new taxa.

In 1928 he moved to Paris and read agricultural bacteriology and serology at the Institut Pasteur. From 1929 to 1930 he studied ichthyology at the Muséum national d'Histoire naturelle under professors L. Roule and J. Pellegrin, as well as carrying out bacteriological and phytopathological investigations in the laboratory of the agricultural firm of G. Truffaut in Versailles. He was elected a 'Correspondant du Musée' and received the gold medal of the Société Centrale d'Aquiculture et de la Pêche, mostly in recognition of his work on Subcarpathian fishes. Vadim remembered seeing Lindberg in Paris when he landed after his trans-Atlantic flight.

In September 1930 he arrived in Canada and began to work as an assistant scientist in biology under Dr. A.G. Huntsman for the Biological Board of Canada (which later became the Fisheries Research Board of Canada). About 1933 or 1934 he became a member of the American Society of Ichthyologists and Herpetologists after visiting Carl L. Hubbs at the University of Michigan. He attended many of their meetings in subsequent years. He worked at the biological stations at St. Andrews, N.B., Halifax, N.S. and at Toronto, Ontario up to 1936. At Toronto with his intriguing accent and his long opera cloak, he was reported to cut a dashing figure with the ladies. However, the life of a beginning immigrant is often hard. He told

Fig. 1. Dimitriy Iv. Vladykov (1863–1955), father ot Vadim D. Vladykov.

Fig. 2. Vadim Vladykov collecting fishes, as a young man in Europe.

a story of waiting for a train, dejected, at a railway station. He did not know much English and had lost the text for a speech he had been asked to make. But seeing a skunk wander by, his interest in life perked up again; always the naturalist. Among other studies during this period he published a major work on the marine fishes of Hudson Bay and a booklet on the marine fishes of Nova Scotia. In 1936 he became a Canadian citizen.

In 1936 after joining the Maryland Conservation Department as fisheries biologist, he undertook studies on shad and striped bass. He also became the first professor of ichthyology at the University of Maryland at College Park and the Biological

Fig. 3. Vadim Vladykov, with co-workers, on steps of biological station in the Parc des Laurentides, May 1941. Left to right: Charles Gauthier, entomologist; V.D. Vladykov; Pierre Lesperance, summer student assistant, now patent lawyer; Vianney Legendre, ichthyologist, now retired but still an active scholar.

Fig. 4. Vadim, 14 May 1934, at Toronto.

Station at Solomons, Maryland. In 1938 he returned to Canada and joined the Laboratoire de Zoologie at the Université de Montréal as professor of ichthyology under the directorship of Dr. Georges Préfontaine. He stayed in this post until 1942. During this period he founded and directed biological stations in the Parc des Laurentides and on Lac St-Louis. Amongst more professional duties he took the Governor General Count Athlone and Her Royal Highness Princess Alice fishing at Grand Lac Jacques Cartier in September 1941. Those knowing Vadim will wonder if he inveigled the Princess into preserving some of her catch for him.

In 1943 Dr. Vladykov organized the Service de Biologie in the Département des Pêcheries du Québec, under the deputy minister Dr. Arthur Labrie, and was its director from 1943 to 1958. During this period he gave courses in ichthyology every two or three years to students in the Ecole Supérieure des Pêcheries at Ste-Anne-de-la-Pocatière. He undertook studies of the fauna of the estuary of the St. Lawrence River and published an important baseline study on the beluga whale that is of considerable value today. He showed that, contrary to fears of commercial fishermen, the beluga was not a significant predator on commercial stocks of fishes. During this period Prof. Vladykov received a proposal from a recent World War II veteran that belugas, now protected, be bombed from the air with 50 pound sacks of potatoes. He also published a series of popular bilingual albums, *The fishes of Quebec,* lithographed in color, on the cod, tomcod, sturgeon, trout, salmon, eel and lobster. He published the most detailed study ever carried out on the osteology of the Pacific salmon after having converted the basement of his own home into a private laboratory containing a unique collection of at least 20000 bones. He built up a scientific collection of fishes (though the latter was not in his assigned duties). Amongst other duties he helped plan and build, under the direction of Dr. Arthur Labrie, assistant Minister of Fisheries, the Aquarium de Québec. Work began on the aquarium in 1955 and it received its first visitors in June 1959. This aquarium developed from several aquaria kept in the offices on rue d'Artigny to please and educate the public about natural resources. It succeeded so much that departmental officials often used the area for semi-official receptions.

Some ichthyologists distance themselves from commercial fishermen and anglers. Vadim, in Nova Scotia and Québec, built relationships with a network of fishermen. They sometimes brought him new records, he listened to them, and quoted them. 'Our old fishermen say wisely, that the lake sturgeon has five kinds of meat. The flesh from various parts of the head and body has a distinct flavour of its own, tender, rich, and always greatly appreciated.'

Vadim was proud of a monster he tracked down and identified in Lac Pohénégamook near St-Eleuthere, in Kamouraska and Témiscouata counties, Québec. Boaters and swimmers on this lake reported sighting large underwater monsters in 1957. Such stories had been circulating for about 50 years. The monster was supposed to be between 12 and 18 feet (4 to 6 m) long, brown or black in color with a rounded back two or three feet wide (60 to 90 cm), and a saw-tooth ridge or fin down the centre. Some saw three humps like a camel, others likened it to a crocodile or sea-cow. One gentleman, who had lost his leg in the war, on sighting the monster paddled furiously to shore and jumped to the safety of land, forgetting his crutches in his haste to escape. Some sought to explain the sightings as explosions of gases from lake sediments, others as a whale, the splashing of a group of mergansers. The worried public contacted the Department of Tourism, Game and Fish, and Vadim investigated, armed with harpoons.

Search for the monster by Vadim and his assistant Marcel Shields was not successful, although many nets were set, the lake was sounded and the outboard engine caught fire. But Vadim was convinced that there was some basis for the stories because there was considerable agreement between the observers several of whom described crenulations or crests on the back. Vadim deduced from the descriptions and behaviour of the beast that it could not be a bird, mammal or reptile. He found that the parish priest brought live sturgeons from the St. Lawrence and kept them tethered in lake tributaries, planning to raise them for their caviar. Vadim hypothesized that some of these sturgeons escaped and grew to large size. The crenulations or crests on the back could correspond to the scutes of a sturgeon. Sturgeons commonly feed in shallow water and sometimes leap partially or entirely out of the water. So it was not unexpected that the sturgeons should be seen, and to the uninitiated might impress the viewer as being a sea monster. In later years several monster watchers and the Société Internationale de Cryptozoologie requested information on the case, to Vadim's delight. He had letters requesting information from as far away as Côte d'Ivoire, Africa.

Vadim was involved with another monster, perhaps the largest great white shark ever caught. In 1936 in *The marine fishes of Nova Scotia* Vladykov and Mackenzie reported 'According to the information of V.D. Vladykov, a 37 foot Man-eater Shark was found dead in a Herring weir on White Head Island (near Grand Manan) in the middle of June 1930. From the liver 210 gallons of oil were obtained'. Vadim had not seen the shark himself but recorded information provided by others. This record was reported in the Guinness Book of World Records. Recent investigations by John Randall of Hawaii's Bishop Museum showed there were as yet no authenticated great white shark longer than 19.5 feet long. However since then a tooth of the White Head Island shark has been found. It authenticates the species as the great white shark, *Carcharodon carcharias*. The tooth size does not permit one to estimate the shark's length as greater than about 16 feet (5 m), according to preliminary study by W.B. Scott of the Huntsman Marine Laboratory. But one cannot eliminate the possibility that the tooth is a small one from near the margin of the jaws of a larger shark. I am searching for other evidence or witnesses.

In 1958 Vadim left the Ministère de Pêcheries du Québec and joined the staff of the University of Ottawa as a professor. There he taught ichthyology, herpetology, zoogeography, etc. and supervised a number of graduate students, S.U. Qadri, C.G. Gruchy, D.T. James, P.K. Liew, C.B. Renaud, J. Lanteigne and B.W. Coad. His educational work extended to the public. He gave numerous lectures to such organizations as the Legion, the YMCA, and to children at the University of Ottawa's Christmas science lectures. Because of his wide knowledge of natural history, and he was one of the few 'whole animal' biologists on staff, he was often asked to identify unusual specimens brought in by students or the public.

During this period he also carried on a number of research projects. He remained in this post until his retirement in 1973. He took one year and three month's absence in Iran starting in August 1961. There he served as a fisheries expert in the Caspian Sea region for FAO. Vadim worked on the sturgeon fisheries, commercially important for caviar. He also made significant collections of fishes, many now at the National Museum of Natural Sciences, Ottawa. He went on cruises of the CSS *Hudson* and *Baffin* to the Caribbean Sea in 1965, 1966 and 1968 to locate the spawning grounds and study the larvae of the American eel. During one of these trips he learned that the ship's itinerary had not yet been determined and resolved to change the situation. He went up to the bridge (where only the crew was allowed) and persuaded the captain to follow a course which suited VDV's plans. The Chief Scientist did not realize what was happening before a few hundred nautical miles had been sailed. However it was the ship's captain that was reprimanded, not Vladykov. In 1972 he became associated with Environmental Control Consultants Ltd., for whom he acted as an advisor for over a decade. Amongst other work he was consulted on were the limnology and fisheries resources of Pukaskwa National Park, the Great Bear River hydroelectric development, and the Champlain-Richelieu River flood control structure.

In 1973 he retired and was named Professor Emeritus of Biology at the University of Ottawa where he retained an office until his death. In 1977 he fulfilled a childhood dream to cross the Soviet Union on the trans-Siberian railroad, and visited Lake Baikal, famous for its endemic cottoid fishes, and Japan. During this trip he visited his older brother in the Ukraine and his younger sister in Belgium, both now deceased. With federal and New Brunswick grants he helped Micmac Indians with a project to raise elvers to sell to the Christmas market in Italy.

Prof. Vladykov felt that high quality illustrations were very important in science. He spent considerable time and effort ensuring that the illustrations showed fine detail. George Ben-Tchavtchavadze, a fellow émigré from Georgia, worked with Vadim to produce superb photos of larval and adult lampreys, cod tubercles and cod gonads, and P.I. Voevodine, also a Russian émigré, made exquisite sepia drawings of skull bones (Fig. 6, also p. 94 in this issue) of Pacific salmon (*Oncorhynchus*).

Nina, his wife, helped in Vadim's researches, especially in the early years. She devoted particular attention to the beluga studies, in research at the Laurentides Park Laboratory, and during their sojourn in Iran.

In later years Prof. Vladykov received a number of honours. In April 1963 he was elected a Fellow of the Royal Society of Canada. In 1968 an entire issue of the Věstník Československé Společnosti Zoologické (see Prelude) was published in honor of his 70th birthday. The Director of the Institute for Vertebrate Research, Czechoslovak Academy of Sciences, Prof. Dr. J. Kratochvil, wrote to him at this time, 'Reading your works we remember that you have laid the main stones of our ichthyology' ... and '... your works are in our country much appreciated ... and utilized'. In 1969 he was elected as The Honorary Member of the Slovak Zoological Society. Following his retirement in 1973 he was appointed Research Associate (2 August) at the National Museum of Natural Sciences, National Museums of Canada. He was asked to give the introductory speech at the American Society of Ichthyologists and Herpetologists annual meeting held in Ottawa, June 1974. In 1975 he was elected an honorary member of the Canadian Society of Zoologists, in November 1976 he received a Doctor of Laws, *honoris causa* degree at Wilfrid Laurier University, Waterloo, and in 1978 he received a D. Sc. *honoris causa* degree from the University of Ottawa. On the 25th anniversary of the Aquarium of Quebec, 29 November 1984, he and Dr. Arthur Labrie were given commemorative plaques for their efforts in creating the aquarium. A wing of the biological station at Point Noir, near the Saguenay Fjord, devoted to the study of whales, was named after him (circa 1985). These honors must have made Vadim, surely at times lonely and somewhat isolated as an immigrant, feel that he had been accepted by his new compatriots. The first volume of *The freshwater fishes of Europe* (edited by Juraj Holčík, 1986), which dealt with lampreys, was dedicated to the memory of Vadim D. Vladykov, 'who made fundamental contributions to our knowledge of taxonomy, systematics, zoogeography and ecology of lampreys'. During this post retirement period up to 1985 he actively carried on research with occasional interruptions because of ill health.

Vladykov was stocky (about 165 cm) and vigorous. He talked with enthusiasm about his current research or he would look at you piercingly with light blue eyes through his thick glasses (for which he was noted) and strongly criticize your work or that of some noted colleague and then crack a joke. He searched for truth and would not excuse the least error. He marshalled those around him, students, colleagues and family, to help him in his projects, sometimes with scant regard for their priorities and sometimes with diplomacy. His students became fearful of extra work when he approached saying 'You, handsome, strong young man ...!'. But many learned how to carry out research under his tutelage.

Fig. 5. Photo of eel-tagging field team on the coast of New Brunswick using an electrical spot welder to weld stainless steel ring tags. Prof. Vladykov is at the left.

His views were strongly held and he would disagree with noted figures of his day such as Carl L. Hubbs on lamprey classification and Johannes Schmidt on the location of the spawning grounds for the American eel. As a splitter he crossed swords with Reeve M. Bailey on generic and specific classification of lampreys and salmonids. In some ways he was unorthodox. He saw a flaw in Ernst Mayr's views about species change. For him there was a conflict between the absolute gap among sympatric species and the relative gap between allopatric species. He felt that the question of species change and adaptation, if not evolution, had not yet been fully answered. In the Quebec Fisheries Department his superior had difficulties restraining Vadim from undertaking studies not related to his duties. From these points one might deduce that Vadim disliked authority. But the reality is, that he fought or argued with almost everyone, student, family, colleagues, administrators (!), from fullness with life and convictions. His criticisms and public addresses were often leavened with humor. In his welcoming speech to the 1974 ASIH meetings in Ottawa Vadim said, 'Personally, I shall treat most cordially all of you, even those who call a lake trout *Salvelinus* instead of *Cristivomer,* or the American brook lamprey as *Lampetra* instead of by the beautiful name of *Lethenteron* proposed by Creaser and Hubbs in 1922'.

He spoke several languages fluently (but all with the famous Vladykov accent). He wrote in several languages: 50% of his papers were in English, 40% in French, 5% in Czech, 3% in German and 2% in Russian (Renaud 1986). His presentations at scientific meetings were salted with humorous comments, delivered in either of Canada's official languages; anglophone and francophone listeners usually assumed that Vadim must speak the *other* language without accent. While a dedicated student of taxonomy, he doubted evolution and was a devout member of the Russian Orthodox Church – he was a founding member of the Church of St. Nicholas in Ottawa, a generous contributor to its memorial fund and to the Russian Orthodox seminary in New York.

VDV had scant regard for ownership of specimens lent to him by museums. He would keep specimens long past their due date. When a student at Michigan I was shown a cabinet half full of empty containers of overdue specimens lent to him. He once embarrassed me by getting me to catalog a specimen he was going to make a type, which was on loan from a California museum; he was caught! Claude B. Renaud has spent over two years sorting out VDV's collections and returning his loans. One loan, a syntype, had been outstanding for 58 years (the BMNH graciously ceded this specimen to NMC).

Persons co-authoring papers with VDV were faced with small and large problems and needed *patience.* Despite the little numbered plastic tags that VDV attached to his important specimens, he was disorganized. He mixed up and misplaced specimens (even holotypes), photos, manuscripts and papers. The disorganization was innate and not just due to his short-sightedness. At times he would be reluctant to give up ideas, even in the face of facts. The co-author or assistants would seemingly convince Vadim of the 'truth', only to have Vadim not so sure the next day. But in the end he would accept. On the other hand, Vadim was very keen in discovering the mistakes of others. Only the meek could avoid arguments with Vadim, but often even the meek were driven to argue with him!

Another Vadim appeared when someone had something that he wanted. It might be lamprey specimens, financial support or a journal in which to publish his paper. He would be diplomatic, flattering or cozening, as need be. And he would be most persistent, there was a bit of the bulldog in him.

Vadim was also a good raconteur. He could hold his audience, child or scientist, spellbound, or make them laugh by teasing on a contested subject. One of his students told me that his stories would so

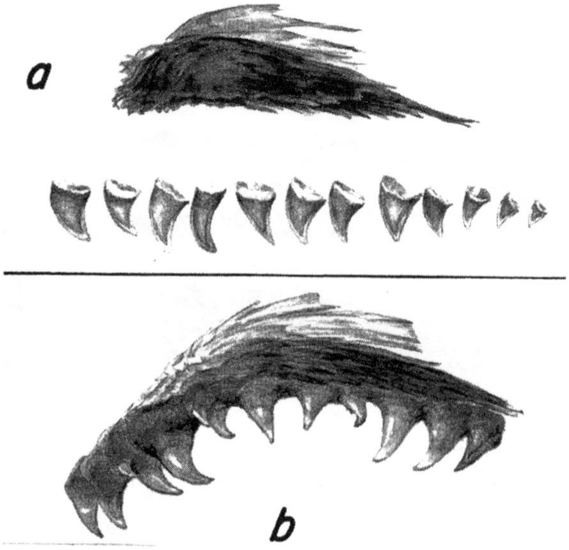

Fig. 6. Sepia drawings of premaxillary bones of pink salmon (*Oncorhynchus gorbuscha*) prepared by P.I. Voevodine under Vladykov's direction, one of a series done for his classic monograph 'Osteological studies on the Pacific salmon of the genus *Oncorhynchus*'. Prof. Vladykov believed strongly in the need for meticulously accurate scientific illustrations.

infuse the preserved lizards with life that they almost seemed to crawl up the wall. He loved art and *objets d'art.* He had a collection of books on painting, he collected carvings in wood and soapstone. He read good literature in Russian, French and English.

Vadim liked the contact with nature that field work brought and he went into the field almost every year. But he also brought nature back to the lab to study and for teaching purposes. This love of nature got him into trouble with the char staff at the University of Ottawa. A mother rattlesnake he kept, gave birth to, he thought, a dozen young. Small enough to escape through the mesh of the mother's cage, they panicked the char staff, who went on strike until all twelve were rounded up. But he had miscounted (see Fig. 8) and later one more young rattler was found roaming around free!

Despite his disputatiousness, his colleagues regarded him with affection, and concurred that he had made major contributions to taxonomy. He was the first one to discover taxonomic characters

Fig. 7. Nina and Vadim Vladykov at the 1973 receptions where he was made Professor Emeritus of the University of Ottawa. Nina assisted Vadim in his researches in Canada and in Iran.

enabling identification of larval lampreys, a discovery of great practical importance when sea lampreys invaded the upper Great Lakes and decimated sport and commercial fisheries. Knowing how to identify larvae saved millions of dollars since no control measures were needed in streams containing only non-parasitic lampreys. Despite imperfect eyesight, he was a keen observer. He was the first to report tubercles on spawning lake charr, a species which had been very well studied. He explored differences in velar tentacles which were useful in lamprey taxonomy and was investigating the value of length of the genital papilla in spawning male lampreys up until a short time before his death. He authored or co-authored (with Prof. E. Kott, W.I. Follett, or students) the first scientific descriptions of more than one quarter of the world's species of lampreys (10 out of 38), more than any other scientist. While some of his colleagues have not accepted some of his generic as-

Fig. 8. The escape of the young and a mother rattlesnake that Prof. Vladykov kept in his lab occasioned this cartoon.

signments, neither have they published refuting evidence. During a career spanning about 63 years (plus 2 years of posthumous publication), he published about 290 scientific papers totalling over 4000 pages. Two other publications were the English translations of papers published previously in French; these translations, not included in the previous page total, comprised 406 pages. He completed an additional five manuscripts, totalling 254 pages.

Vladykov contributed significantly to ichthyology and fish biology in Canada. He carried out pioneering ichthyological studies when very little was being done. He discovered many taxa new to science. He established several biological stations in Canada and trained their staff. He taught courses and supervised several graduate students. Despite being uprooted early in his life, by the force of his personality, his capacity to question dogma and to work hard, the keenness of his observation, and his numerous publications, Professor Vladykov significantly influenced the development of fish biology and ichthyology in Canada and made his mark in international ichthyology.

Fig. 9. Vadim Vladykov, photo in late 1986. Prof. Vladykov published at least 290 scientific papers in his life, totalling over 4000 pages. Note the resemblance to his father in Figure 1. Photograph by D.E.M.

Acknowledgements

The author is grateful to George Ben-Tchavtchavadze, E.W. Burridge, François Chapleau, Brian W. Coad, C.G. Gruchy, J. Holčík, E. Kott, C.C. Lindsey, John McNeill, Claude B. Renaud, and W.B. Scott for information or comments on this manuscript.

Sources

Besides various *curriculum vitae* prepared by Dr. Vladykov at several stages of his career, information drawn from his publications and his friends, I used the following documents:

Dodson, E.O. 1973. Notes for a speech on Vladykov's retirement from the University of Ottawa. 4 pp.

Plamondon, M. 1959. Dr. Vadim D. Vladykov, Ph. D. Actualités Marines 3: 22–27 (illus.).

Pawlick, T. 1977. The Pohénégamook enigma, Quebec's answer to the Loch Ness monster. Harrowsmith: 25–33, 110.

Renaud, C.B. 1986. Obituary of Vadim Dimitrij Vladykov. Copeia 1986: 559–560.

Appendix 1

List of new taxa authored by V.D. Vladykov.

The following new taxa were named by Vladykov or by Vladykov and his co-authors between 1925 and 1982. This includes the following numbers of new taxa: 1 genus, 2 subgenera, 17 species, 10 subspecies, 5 natios, and 1 morpha.

Gobio gobio carpathicus n. subsp. Vladykov, 1925
Gobio Frici n. sp. Vladykov, 1925
Gobio uranoscopus carpathorossicus n. subsp. Vladykov, 1925
Lampetra Bergi n. sp. Vladykov, 1925
Cobitis montana n. g. Vladykov, 1925
Sabanejewia n. g. Vladykov, 1929
Rutilus rutilus carpathorossicus n. subsp. Vladykov, 1930
Rutilus aula natio *Karamani* nat. n. Vladykov & Petit, 1930
Leuciscus svallize zrmanjae natio *Risae* nat. n. Vladykov & Petit, 1930
Scardinius scardafa natio *ohridana* nat. n. Vladykov & Petit, 1930
Cottus gobio natio *Pellegrini* nat. n. Vladykov, 1931
Leuciscus leuciscus natio *Roulei* nat. n. Vladykov, 1931
Lumpenus lampetraeformis americanus n. subsp. Vladykov, 1935
Lumpenus lampetraeformis terrae-novae n. subsp. Vladykov, 1935
Lycodes atratus n. sp. Vladykov & Tremblay, 1936
Lycodes lavalaei n. sp. Vladykov & Tremblay, 1936
Lycodes reticulatus laurentianus n. subsp. Vladykov & Tremblay, 1936
Lycodes turneri atlanticus n. subsp. Vladykov & Tremblay, 1936
Lycodes turneri atlanticus morpha *squamosa* n. morpha Vladykov & Tremblay, 1936
Lycodes vachonii n. sp. Vladykov & Tremblay, 1936
Lycodes vahlii n. sp. Vladykov & Tremblay, 1936
Lampetra zanandreai n. sp. Vladykov, 1955
Acipenser oxyrhynchus de sotoi n. subsp. Vladykov, 1955
Salvelinus marstoni intermedius n. subsp. Vladykov, 1957
Salvelinus marstoni cavanaghi n. subsp. Vladykov, 1957
Melanostigma normani Vladykov *nomen nudum* in Vladykov (and McAllister) 1961
Parasalmo n. subg. unavailable Vladykov, 1963
Parahucho n. subg. unavailable Vladykov, 1963
Lampetra richardsoni n. sp. Vladykov & Follett, 1965
Parahucho n. subg. available Vladykov in Vladykov & Gruchy, 1972
Parasalmo n. subg. available Vladykov in Vladykov & Gruchy, 1972
Lampetra pacifica n. sp. Vladykov, 1973
Lethenteron meridionale n. sp. Vladykov, Kott & Pharand-Coad, 1975
Entosphenus folletti n. sp. Vladykov & Kott, 1976
Entosphenus hubbsi n. sp. Vladykov & Kott, 1976
Lethenteron alaskense n. sp. Vladykov & Kott, 1978
Lethenteron matsubarai n. sp. Vladykov & Kott, 1978
Entosphenus similis n. sp. Vladykov & Kott, 1979
Eudontomyzon hellenicus n. sp. Vladykov, Renaud, Kott & Economidis, 1982.

Environmental Biology of Fishes Vol. 23, No. 1–2, pp. 21–36, 1988

Bibliography of Vadim D. Vladykov: scientific publications and manuscript reports 1923–1988

Brian W. Coad, Don E. McAllister & Claude B. Renaud
Ichthyology Section, National Museum of Natural Sciences, P.O. Box 3443, Station D, Ottawa, Ontario, Canada K1P 6P4

V.D. Vladykov produced over 290 papers and unpublished manuscript reports in a writing career spanning the 64 years from 1923 to 1986, an average of about 4 publications per year. A few of these consisted only of titles of papers read at meetings. They ranged in length from a few lines to classic monographs and totalled over 4000 pages. Four hundred and six additional pages were published as translations from the original. His works ranged in topic from the biology of whales to the diseases and parasites of cereals but the bulk of his work centered on fishes with a systematic bias and a style peculiar to 'VDV'. These varied articles were published in several languages, namely French (109), Czech (11), German (6), Ruthenian (3), bilingual English/French (7) and Russian (1), with the bulk in English. He shared authorship with 39 collaborators 90 times but he was first author on 81% of these!

Where we have had access to the publication we have attempted to supply the number as well as the volume of the journal (where applicable), and the number or lack of illustrations. We feel a bibliography should be more complete than References Cited. Papers read at the annual meetings of the American Society of Ichthyologists and Herpetologists are cited from reports of these meetings in Copeia rather than from programme abstracts which were only available to those attending the meetings. Citations are arranged by year of publication but we have made no effort to arrange them by date within years. We have included a number of manuscript reports which we felt to be significant. Generally however, we have omitted manuscript reports when they were later published virtually unchanged. When in possession of the date of publication of new taxa we included it.

1. * Vladykov, V.D. 1923. Fisheries of the Kertch Channel. Bull. of Institute for Study of Russia, Prague 1: ?–?. (In Russian).

2–7. * Vladykov, V.D. 1925–1928. Six articles on the fishes and fisheries of Subcarpathia. Rybařský Věstník, Prague. (In Czech).

8–9. * Vladykov, V.D. 1925. Two ichthyological notes on fishes from Subcarpathia. Věda Přírodní, Prague. (In Czech).

10. Vladykov, Vadim. 1925. Über einige neue Fische aus der Tschechoslowakei (Karpathorussland). Zool. Anzeiger 64(11/12): 248–252, 0 fig. [published 20 November 1925; *Gobio gobio carpathicus* n. subsp., *Gobio Frici* n. sp., *Gobio uranoscopus carpathorossicus* n. subsp., *Lampetra Bergi* n. sp.].

11. Vladykov, Vadim. 1925. Über eine neue Cobitis-Art aus der Tschechoslovakei: Cobitis montana n. sp. Zool. Jahrb. Abt. f. System., Geogr. Biol. Tiere 50: 320–338, text-fig. A & B, 9 fig. [published ?; *Cobitis montana* n. sp.].

12. Vladykov, Vadim. 1925. Dimorphisme sexuel chez la loche de rivière (Cobitis taenia taenia L.). Věstnik Kral. Čes. Spol. Nauk. Tr., Prague 2: 1–8, 2 fig. (In Czech and French).

13. Vladykov, Vadim. 1926. Topography and histology of the lateral line system in brown trout (Trutta fario L.). Public. Faculté Sci. Univ.

* = references which have not been verified against a copy of the publication.

[] = following the references we have noted, enclosed in brackets, new taxa which are described and, where we have been able to determine it, a more precise date of publication.

Charles, Prague 57: 1–38, 11 fig., 1 pl. (In Czech; French résumé).
14. Vladykov, Vadim. 1926. Fishes of Subcarpathian Russia and methods of their capture. Uzhorod, Czechoslovakia. v + 147 pp., 75 fig., 1 map. (In Ruthenian).
15. Vladykov, Vadim. 1927. Common names for fishes from various localities of Subcarpathian Russia. Recueil scientifique 'Prosvita', Uzhorod, Czechoslovakia 5: 205–232, 0 fig. (In Ruthenian).
16. Vladykov, Vadim. 1927. Zur Frage über das Absterben unserer Lampreten (Petromyzonidae) nach der Laichzeit. Zool. Anzeiger 74(7/10): 237–239, 0 fig.
17. Vladykov, Vadim. 1927. Über den geschlechtlichen Dimorphismus bei Ellritzen *Phoxinus phoxinus* (Linné). Zool. Anzeiger 74(11/12): 322–328, 2 fig.
18. Vladykov, Vadim. 1927. Diseases and parasites of cereals from Subcarpathian Russia. Biblioteka Zemledilskogo Referata, Uzhorod, Czechoslovakia 7: 1–95, 59 fig. (In Ruthenian).
19. Vladykov, Vadim. 1927. Mystery of the fauna of Lake Baikal. Přiroda, Brno, Czechoslovakia 20(2/3): 43–48, 0 fig. (In Czech).
20. Pellegrin, Jacques & V. Vladikov [sic]. 1928. Sur les poissons du lac Baïkal appartenant au genre *Cottus*. Bull. Mus. Hist. Nat., Paris 34: 316–318, 0 fig.
21. Vladykov, Vadim. 1928. Über sekundären Geschlechtsdimorphismus bei unseren Cobitiden. Zool. Jahrb. Abt. f. System., Ökol. Geogr. Tiere 55(1/3): 147–162, 6 fig.
22. Vladykov, V. 1929. Sur un nouvel hybride entre le nase (*Chondrostoma nasus* L.) et le zerte (*Vimba vimba* Fitz.). Bull. Soc. Centr. Aquicul. Pêch., Paris 36(10/12): 118–121, 0 fig.
23. Vladykov, V. 1929. Review of: V.J. Stépan. La carpe de Wittingau dégénère-t-elle en Tchécoslovaquie? Bull. Soc. Centr. Aquicul. Pêch., Paris 36(10/12): 125, 0 fig.
24. Vladykov, Vadim. 1929. Sur un nouveau genre de Cobitidés: *Sabanejewia*. Bull. Mus. Hist. Nat., Paris, ser. 2, 1(1): 85–90, 0 fig. [published ?; *Sabanejewia* n. g.].
25–28. * Vladykov, V.D. 1929–1930. Four articles. Jardinage, Versailles.
29. Petit, G. & V. Vladykov. 1930. A propos des salmonidés du Lac d'Ohrida. Bull. Mus. Hist. Nat., Paris, ser. 2, 2(5): 516–518, 0 fig.
30. * Vladykov, V.D. 1930. Ueber die hystologischen Veränderungen in den Muskeln einiger gefrorener im Kühlhause Vaugirard in Paris aufbewahrter Fische. Die Kälte Industrie, Hamburg 27: 13–16.
31. Vladykov, Vadim. 1930. Sur une nouvelle forme de gardon de la Russie sous-carpathique (Bassin du Danube) *Rutilus rutilus carpathorossicus,* subsp. n. Bull. Soc. Zool. France, Paris 55: 103–107, 0 fig. [published 11 February 1930; *Rutilus rutilus carpathorossicus* n. subsp.].
32. Vladykov, Vadim. 1930. Sur l'espèce et ses unités taxonomiques en ichthyologie. Bull. Mus. Hist. Nat., Paris, ser. 2, 2(1): 75–83, 0 fig.
33. Vladykov, V. & G. Petit. 1930. Sur quelques poissons d'eau douce d'Albanie. Bull. Soc. Zool. France, Paris 55: 383–409, 2 fig. [published 10 June 1930; *Rutilus aula* natio *Karamani* nat. n., *Leuciscus svallize zrmanjae* natio *Risae* nat. n., *Scardinius scardafa* natio *ohridana* nat. n.].
34. * Truffaut, Georges & V. Vladykov. 1930. La microflore de la rhizosphère du blé. C.R. Acad. Sci., Paris 190: 824–826, 0 fig.
35. Monod, Th. & V. Vladykov. 1931. Sur quelques copépodes parasites provenant de la Russie sous-carpathique (Tchécoslovaquie). Ann. Parasit. Humaine Compar., Paris 9(3): 202–224, 11 fig.
36. Vladykov, Vadim. 1931. Poissons de la Russie sous-carpathique (Tchécoslovaquie). Mém. Soc. Zool. France 29(4): 217–374, 29 fig., 1 map. [published 21 September 1931; *Leuciscus leuciscus* natio *Roulei* nat. n., *Cottus gobio* natio *Pellegrini* nat. n.].
37. Vladykov, V.D. 1932. Fishes from the Hudson bay region. Pollock fishery in the Fundy area. Halibut fishery in the Fundy area. Ann. Rep. Biol. Board Can. (1931): 45–46, 0 fig.
38. Vladykov, Vadim D. 1933. Biological and

oceanographic conditions in Hudson Bay 9. Fishes from the Hudson Bay region (except the Coregonidae). Contrib. Can. Biol. Fish., New Ser. 8(2), Ser. A, General 29: 14–61, 5 fig.

39. Vladykov, V.D. 1933. Seasonal distribution of the haddock of the Halifax area. Young haddock in the Halifax area. Ann. Rep. Biol. Board Can. (1932): 41–42, 0 fig.
40. Vladykov, V.D. 1933. High temperature stops haddock fishing. Progr. Rep. Atl. Biol. Stat., St. Andrew's, N.B. 29: 10–11, 1 fig.
41. Dymond, John R. & V.D. Vladykov. 1934. The distribution and relationship of the salmonoid fishes of North America and North Asia. Proc. Fifth Pacific Sci. Congr., Victoria and Vancouver, 1933, Vol. B6(23): 3740–3750, 10 fig.
42. Vladykov, V.D. 1934. Inshore spring fishery of haddock in the Halifax area. Haddock races along the North Atlantic coast. Data on occurrence and spawning of fishes on outer Nova Scotian bank. Ann. Rep. Biol. Board Can. (1933): 39–41, 0 fig.
43. Vladykov, V.D. 1934. On the occurrence of some North Pacific fishes in Hudson Bay with remarks on the fish fauna of the latter area. Proc. Fifth Pacific Sci. Congr., Victoria and Vancouver, 1933, Vol. B6(29): 3787–3789, 0 fig.
44. Vladykov, V.D. 1934. Young haddock in the vicinity of Halifax, N.S. Contr. Can. Biol. Fish., New Ser. 8(29), Ser. A, General 44: 411–419, 3 fig.
45. Vladykov, V.D. 1934. Geographical variation in the number of rows of pharyngeal teeth in cyprinid genera. Copeia (3): 134–136, 0 fig.
46. Vladykov, Vadim D. 1934. Environmental [sic] and taxonomic characters of fishes. Trans. Roy. Can. Inst., Toronto 20(1): 99–140, 0 fig.
47. Vladykof [sic], V.D. 1934. The value of aquaria. Animal Life, Toronto, 23(3): 36, 0 fig.
48. Vladykov, V.D. 1935. A 'Three-eyed' haddock (*Melanogrammus aeglefinus*) caught at St. Andrews, N.B. Can. Field-Nat. 49(4): 64–67, 2 fig.
49. Vladykov, Vadim Dmitrij. 1935. Secondary sexual dimorphism in some Chinese cobitid fishes. J. Morphol. 57(1): 275–302, 8 fig.
50. Vladykov, V.D. 1935. Haddock races along the North American coast. Progr. Rep. Atl. Biol. Stat., St. Andrew's, N.B. 43: 3–7, 2 fig.
51. * Vladykov, V.D. 1935. Haddock races along the North American coast. Canadian Fisherman 22(?): 15–17, ? fig.
52. Vladykov, V.D. 1935. The 'three-eyed' haddock fake. Fish live after operation. Canadian Fisherman 22(8): 14–16, 1 fig.
53. Vladykov, V.D. 1935. Two new subspecies of *Lumpenus lampetraeformis (Walbaum)* from North America. Ann. Rep. Fish. Res. Lab. Newfld. 1934, Appendix C, 3: 75–78, 0 fig. [published March 1935; *Lumpenus lampetraeformis americanus* n. subsp. and *Lumpenus lampetraeformis terrae-novae* n. subsp.].
54. Vladykov, V.D. 1935. Some unreported and rare fishes for the coast of Nova Scotia. Proc. Nova Scotian Inst. Sci. 19 (1934–1935), Part I: 1–8, 0 fig.
55. Vladykov, V.D. & R.E.S. Homans. 1935. Do haddock feed during the spawning period? Progr. Rep. Atl. Biol. Stat., St. Andrew's, N.B. 45: 10–11, 1 fig.
56. Vladykov, V.D. & R.A. McKenzie. 1935. The marine fishes of Nova Scotia. Proc. Nova Scotian Inst. Sci. 19, Part 1: 17–113, 130 fig.
57. Vladykov, V.-D. [sic] & J.-L. Tremblay. 1935. Liste des poissons recueillis pendant l'été 1934 par la Station biologique du St-Laurent, dans la région de Trois-Pistoles, P.Q. Naturaliste can. 62: 77–82, 0 fig. (Contr. Stat. biol. St-Laurent, Trois-Pistoles, Québec, Troisième Rapport, 1934–1935, No. 1).
58. Vladykov, Vadim D. 1936. Occurrence of three species of anadromous fishes on the Nova Scotian Banks during 1935 and 1936. Copeia (3): 168, 0 fig.
59. Vladykov, V.D. 1936. Capsules d'oeufs de raies de l'Atlantique canadien appartenant au genre *Raja*. Naturaliste can. 63: 211–231, 7 fig. (Contr. Stat. biol. St-Laurent, Trois-Pistoles, Québec (6)).

60. Vladikov [sic], V.D. 1936. How to properly keep fish. Nova Scotia Fish and Game (Fishermen's Number), Halifax, May 1936, p. 19, 0 fig.
61. Vladykov, V.D. & J.-L. Tremblay. 1936. Nouvelles espèces de *Lycodes* (Pisces, Zoarcidae) du Saint-Laurent et revision de toutes les espèces du même genre de l'Atlantique occidental. Fauna et Flora Laurentianae, Québec 1: 1–45, 17 fig., 3 tabs. [published 22 April 1936; *Lycodes atratus* n. sp., *Lycodes lavalaei* n. sp., *Lycodes reticulatus laurentianus* n. subsp., *Lycodes turneri atlanticus* n. subsp., *Lycodes turneri atlanticus* morpha *squamosa* n. morpha, *Lycodes vachonii* n. sp., *Lycodes vahlii maculatus* n. subsp.].
62. Truitt, R.V. & V.D. Vladykov. 1937. Striped bass investigations in the Chesapeake Bay. Trans. Amer. Fish. Soc. 66(1936): 225–226, 0 fig.
63. Truitt, R.V. & V.D. Vladykov. 1937. The importance of sport fishing in Maryland. Trans. Amer. Fish. Soc. 66(1936): 403–405, 0 fig.
64. McCurdy, G.A. & V.D. Vladykov. 1938. Differentiation of fresh haddock fillets from cod by the precipitation test. Proc. Nova Scotian Inst. Sci. 19(1936–1937), Part 3: 237–250, 0 fig.
65. Vladykov, Vadim D. 1938. Migration des bars d'Amérique (Roccus lineatus) dans la baie de Chesapeake, établies par l'étiquetage de 1936–37. Annales de l'ACFAS 4: 97, 0 fig.
66. * Vladykov, V.D. 1938. Rapport sur le Marsouin blanc ou Béluga (*Delphinapterus leucas*) de la Province de Québec. Dept. Pêch. maritimes, Québec. 25 pp., ? fig.
67. Vladykov, V.D. & D.H. Wallace. 1938. Remarks on populations of the shad (*Alosa sapidissima*) along the Atlantic coast region. Trans. Amer. Fish. Soc. 67(1937): 52–66, 0 fig. (Contr. Chesapeake Biol. Lab. (17)).
68. Vladykov, V.D. & D.H. Wallace. 1938. Is the striped bass (*Roccus lineatus*) of Chesapeake Bay a migratory fish? Trans. Amer. Fish. Soc. 67(1937): 67–86, 1 fig. (Contr. Chesapeake Biol. Lab. (18)).
69. * Vladykov, V.D. 1939. Rapport préliminaire sur le marsouin blanc ou Béluga (*Delphinapterus leucas*) sur le banc de Manicouagan en 1938. Dept. Pêch. maritimes, Québec. 50 pp., ? fig.
70. Vladykov, V.D. 1939. Recherches sur le marsouin blanc ou béluga (Delphinapterus leucas) dans le Saint-Laurent. Annales de l'ACFAS 5: 101, 0 fig.
71. Bonin, W. & V.D. Vladykov. 1940. Études sur les mammifères aquatiques. I. La peau du Marsouin blanc ou Béluga (*Delphinapterus leucas*). Naturaliste can. 67(10/11): 253–287, 18 fig. (Contr. Dépt. Pêch. Québec (6) and Contr. Inst. Zool., Univ. Montréal (7)).
72. Vladykov, Vadim D. 1940. 'Homing Instinct' de la truite mouchetée (Salvelinus fontinalis). Annales de l'ACFAS 6: 120, 0 fig.
73. Vladykov, Vadim D. 1940. Remarques préliminaires sur la nourriture de la truite mouchetée (Salvelinus fontinalis) du Parc des Laurentides. Annales de l'ACFAS 6: 120, 0 fig.
74. Vladykov, Vadim D. 1940. Nourriture du Marsouin blanc sur les bancs de Manicouagan. Annales de l'ACFAS 6: 120–121, 0 fig.
75. Vladykov, Vadim D. 1940. Remarques sur les cétacés du Saint-Laurent. Annales de l'ACFAS 6: 121, 0 fig.
76. Vladykov, V.D. 1940. A peculiar structure of the anal fin of capelin. Copeia (3): 214 (title only).
77. Vladykov, V.D. 1940. Remarks on the Salvelinus marstoni in the Laurentides Park. Copeia (3): 214 (title only).
78. Vladykov, Vadim D. & Vianney Legendre. 1940. The determination of the number of eggs in ovaries of brook trout (*Salvelinus fontinalis*). Copeia (4): 218–220, 0 fig.
79. Bernier, Germaine & V.D. Vladykov. 1941. Un Trichoptère qui vit sur l'éponge, Spongilla lacustris, dans le parc des Laurentides. Annales de l'ACFAS 7: 118–119, 0 fig.
80. Vladykov, V.D. 1941. Remarques sur la truite rouge du Québec (*Salvelinus marstoni*) dans le parc des Laurentides. Annales de l'ACFAS 7: 119, 0 fig.

81. Vladykov, V.D. 1941. Liste des poissons du parc des Laurentides. Annales de l'ACFAS 7: 120, 0 fig.
82. Vladykov, V.D. 1941. Batraciens du parc des Laurentides. Annales de l'ACFAS 7: 120–121, 0 fig.
83. Vladykov, Vadim D. 1941. The structure of the anal fin in the caplin, *Mallotus,* Osmeridae. Naturaliste can. 68(8/9): 193–207, 3 fig. (Contr. Dépt. Pêch. Québec (8)).
84. Vladykov, Vadim D. 1941. Preliminary list of amphibians from the Laurentides Park in the Province of Quebec. Can. Field-Nat. 55(6): 83–84, 0 fig.
85. Vladykov, Vadim D. 1941. Introduction. *In:* Rapport de la Station biologique de Montréal et de la Station biologique du Parc des Laurentides pour l'année 1941. Travaux du Ministère de la Chasse et de la Pêche de Québec, Institut de Biologie, Université de Montréal, Fascicule 3: 318–323, 0 fig.
86. Vladykov, Vadim D. 1941. Distribution des frayères d'achigans et de dorés dans le lac Saint-Louis. *In:* Rapport de la Station biologique de Montréal et de la Station biologique du Parc des Laurentides pour l'année 1941. Travaux du Ministère de la Chasse et de la Pêche de Québec, Institut de Biologie, Université de Montréal, Fascicule 3: 343–352, 0 fig.
87. Vladykov, Vadim D. 1941. Mouvements des achigans dans le Lac Saint-Louis, d'après l'etiquetage, 1941. *In:* Rapport de la Station biologique de Montréal et de la Station biologique du Parc des Laurentides pour l'année 1941. Travaux du Ministère de la Chasse et de la Pêche de Québec, Institut de Biologie, Université de Montréal, Fascicule 3: 357–368, 0 fig.
88. Vladykov, Vadim D. 1941. Observations sur les 'carpes' (Catostomidés) dans la rivière Châteauguay, 1941. *In:* Rapport de la Station biologique de Montréal et de la Station biologique du Parc des Laurentides pour l'année 1941. Travaux du Ministère de la Chasse et de la Pêche de Québec, Institut de Biologie, Université de Montréal, Fascicule 3: 369–375, 0 fig.
89. Vladykov, Vadim D. 1941. Nouveau poisson pour le Canada, *Ammocrypta pellucida* (Ethéostomidés). *In:* Rapport de la Station biologique de Montréal et de la Station biologique du Parc des Laurentides pour l'année 1941. Travaux du Ministère de la Chasse et de la Pêche de Québec, Institut de Biologie, Université de Montréal, Fascicule 3: 393–394, 0 fig.
90. Vladykov, Vadim D. 1941. Remarques sur la biologie du 'scorpion', *Corydalis cornuta,* dans la rivière Châteauguay. *In:* Rapport de la Station biologique de Montréal et de la Station biologique du Parc des Laurentides pour l'année 1941. Travaux du Ministère de la Chasse et de la Pêche de Québec, Institut de Biologie, Université de Montréal, Fascicule 3: 395–396, 0 fig.
91. Vladykov, Vadim D. 1941. Introduction. *In:* Rapport de la Station biologique de Montréal et de la Station biologique du Parc des Laurentides pour l'année 1941. Travaux du Ministère de la Chasse et de la Pêche de Québec, Institut de Biologie, Université de Montréal, Fascicule 4: 398–406, 0 fig.
92. Vladykov, Vadim D. 1941. Statistiques de pêche à la truite dans le Parc des Laurentides, année 1938. *In:* Rapport de la Station biologique de Montréal et de la Station biologique du Parc des Laurentides pour l'année 1941. Travaux du Ministère de la Chasse et de la Pêche de Québec, Institut de Biologie, Université de Montréal, Fascicule 4: 407–413, 0 fig.
93. Vladykov, Vadim D. 1941. Rapport entre la longueur et le poids chez la truite du Parc des Laurentides, 1938–1939–1940. *In:* Rapport de la Station biologique de Montréal et de la Station biologique du Parc des Laurentides pour l'année 1941. Travaux du Ministère de la Chasse et de la Pêche de Québec, Institut de Biologie, Université de Montréal, Fascicule 4: 414–420, 0 fig.
94. Vladykov, Vadim D. 1941. Conditions de pêche à la truite dans le camp Jacques-Cartier, durant 1936–40. *In:* Rapport de la Station biologique de Montréal et de la Station

biologique du Parc des Laurentides pour l'année 1941. Travaux du Ministère de la Chasse et de la Pêche de Québec, Institut de Biologie, Université de Montréal, Fascicule 4: 421–442, 0 fig.

95. Vladykov, Vadim D. 1941. Productivité sportive en truites de quelques lacs du Parc des Laurentides. *In:* Rapport de la Station biologique de Montréal et de la Station biologique du Parc des Laurentides pour l'année 1941. Travaux du Ministère de la Chasse et de la Pêche de Québec, Institut de Biologie, Université de Montréal, Fascicule 4: 443–446, 0 fig.
96. Vladykov, Vadim D. 1941. Nature de la population des truites dans le Parc des Laurentides. *In:* Rapport de la Station biologique de Montréal et de la Station biologique du Parc des Laurentides pour l'année 1941. Travaux du Ministère de la Chasse et de la Pêche de Québec, Institut de Biologie, Université de Montréal, Fascicule 4: 470–480, 0 fig.
97. Vladykov, Vadim D. & Charles Gauthier. 1941. Régime alimentaire des jeunes achigans (*Micropterus dolomieui*) dans la région de Montréal. *In:* Rapport de la Station biologique de Montréal et de la Station biologique du Parc des Laurentides pour l'année 1941. Travaux du Ministère de la Chasse et de la Pêche de Québec, Institut de Biologie, Université de Montréal, Fascicule 3: 376–380, 0 fig.
98. Vladykov, Vadim D. & Charles Gauthier. 1941. Régime alimentaire des jeunes dorés (*Stizostedion vitreum*) dans la région de Montréal. *In:* Rapport de la Station biologique de Montréal et de la Station biologique du Parc des Laurentides pour l'année 1941. Travaux du Ministère de la Chasse et de la Pêche de Québec, Institut de Biologie, Université de Montréal, Fascicule 3: 381–383, 0 fig.
99. Vladykov, Vadim D. & Charles Gauthier. 1941. Remarques sur le régime alimentaire de l'esturgeon (*Acipenser fulvescens*) dans le Lac Saint-Louis. *In:* Rapport de la Station biologique de Montréal et de la Station biologique du Parc des Laurentides pour l'année 1941. Travaux du Ministère de la Chasse et de la Pêche de Québec, Institut de Biologie, Université de Montréal, Fascicule 3: 384–387, 0 fig.
100. Vladykov, Vadim D. & Charles Gauthier. 1941. Remarques sur le régime alimentaire du ouaouaron (*Rana catesbiana*) dans la rivière Châteauguay. *In:* Rapport de la Station biologique de Montréal et de la Station biologique du Parc des Laurentides pour l'année 1941. Travaux du Ministère de la Chasse et de la Pêche de Québec, Institut de Biologie, Université de Montréal, Fascicule 3: 388–390, 0 fig.
101. Vladykov, Vadim D. & Charles Gauthier. 1941. Relation entre la taille des alevins de truites et la productivité des ruisseaux en nourriture, dans le Parc des Laurentides. *In:* Rapport de la Station biologique de Montréal et de la Station biologique du Parc des Laurentides pour l'année 1941. Travaux du Ministère de la Chasse et de la Pêche de Québec, Institut de Biologie, Université de Montréal, Fascicule 4: 462–469, 0 fig.
102. Vladykov, V.D. & Charles-Henri Gauthier. 1941. La truite mouchetée (Salvelinus fontinalis) comme collectionneur d'insectes terrestres. Annales de l'ACFAS 7: 119, 0 fig.
103. Vladykov, V.D. & Vianney Legendre. 1941. La résistance de la truite mouchetée (Salvelinus fontinalis) à différentes teneurs en O_2 et CO_2. Annales de l'ACFAS 7: 122, 0 fig.
104. Vladykov, V.D. & Vianney Legendre. 1941. Hydrographie des lacs du parc des Laurentides. Annales de l'ACFAS 7: 122–123, 0 fig.
105. Vladykov, Vadim D. & Vianney Legendre. 1941. Remarques sur l'hydrographie du Lac Saint-Louis. *In:* Rapport de la Station biologique de Montréal et de la Station biologique du Parc des Laurentides pour l'année 1941. Travaux du Ministère de la Chasse et de la Pêche de Québec, Institut de Biologie, Université de Montréal, Fascicule 3: 324–336, 0 fig.
106. Vladykov, Vadim D. & Vianney Legendre. 1941. Liste des poissons observés dans la rivière Châteauguay, et leurs noms locaux. *In:*

Rapport de la Station biologique de Montréal et de la Station biologique du Parc des Laurentides pour l'année 1941. Travaux du Ministère de la Chasse et de la Pêche de Québec, Institut de Biologie, Université de Montréal, Fascicule 3: 337–342, 0 fig.

107. Vladykov, Vadim D. & Vianney Legendre. 1941. Etiquetage des poissons autour de Montréal, durant 1941. *In:* Rapport de la Station biologique de Montréal et de la Station biologique du Parc des Laurentides pour l'année 1941. Travaux du Ministère de la Chasse et de la Pêche de Québec, Institut de Biologie, Université de Montréal, Fascicule 3: 353–356, 0 fig.

108. Vladykov, Vadim D. & Vianney Legendre. 1941. Etiquetage des brochets (*Esox lucius*) dans le Lac Saint-Louis, 1941. *In:* Rapport de la Station biologique de Montréal et de la Station biologique du Parc des Laurentides pour l'année 1941. Travaux du Ministère de la Chasse et de la Pêche de Québec, Institut de Biologie, Université de Montréal, Fascicule 3: 391–392, 0 fig.

109. Vladykov, Vadim D. & Vianney Legendre. 1941. Observations physico-chimiques sur les lacs du Parc des Laurentides de bonne heure au printemps. *In:* Rapport de la Station biologique de Montréal et de la Station biologique du Parc des Laurentides pour l'année 1941. Travaux du Ministère de la Chasse et de la Pêche de Québec, Institut de Biologie, Université de Montréal, Fascicule 4: 447–451, 0 fig.

110. Vladykov, Vadim D. & Vianney Legendre. 1941. Conditions physico-chimiques durant la fraye des truites mouchetées dans les eaux du Parc des Laurentides, 1938–39–40–41. *In:* Rapport de la Station biologique de Montréal et de la Station biologique du Parc des Laurentides pour l'année 1941. Travaux du Ministère de la Chasse et de la Pêche de Québec, Institut de Biologie, Université de Montréal, Fascicule 4: 452–461, 0 fig.

111. Vladykov, V.D., V. Legendre & J.-P. Cuerrier. 1941. Variations régionales du nombre des oeufs chez la truite mouchetée (Salvelinus fontinalis) du parc des Laurentides. Annales de l'ACFAS 7: 121–122, 0 fig.

112. Vladykov, V.D. & Émile Rochon. 1941. Liste préliminaire des Odonates du Parc des Laurentides. Annales de l'ACFAS 7: 121, 0 fig.

113. Vladykov, Vadim D. 1942. Étude des lacs du Parc des Laurentides, 1938–1941. Volume I. Rapport au Ministère de la Chasse et des Pêcheries, Province de Québec. 75 pp., 0 fig. [Revised and reproduced in 1947].

114. Vladykov, Vadim D. 1942. Étude des lacs du Parc des Laurentides, 1938–1941. Volume II. Conditions physico-chimiques et biologiques, et statistiques de la pêche dans 48 lacs du Parc des Laurentides. Rapport au Ministère de la Chasse et des Pêcheries, Province de Québec. 178 pp., 0 fig. [Revised and reproduced in 1949].

115. Vladykov, Vadim D. 1942. Remarques sur la biologie du 'scorpion', Corydalis cornuta, dans la rivière Châteauguay. Annales de l'ACFAS 8: 109, 0 fig.

116. Vladykov, Vadim D. 1942. Nouveau poisson pour le Canada, Ammocrypta pellucida, Éthéostomidés. Annales de l'ACFAS 8: 111, 0 fig.

117. Vladykov, Vadim D. 1942. Precision with which speckled trout (Salvelinus fontinalis) return to the same spawning grounds. Can. Field-Nat. 56(8/9): 134–136, 1 fig.

118. Vladykov, Vadim D. 1942. Two fresh-water fishes new for Quebec. Copeia (3): 193–194, 0 fig.

119. Vladykov, Vadim D. & Charles Gauthier. 1942. Nourriture des jeunes achigans (Micropterus dolomieui) dans la région de Montréal. Annales de l'ACFAS 8: 110, 0 fig.

120. Vladykov, Vadim D. 1943. Studies on aquatic mammals. II. A modification of the pectoral fins in the Beluga from the St. Lawrence River. Naturaliste can. 70(1/2): 23–40, 7 fig. (Contr. Dépt. Pêch. Québec (10)).

121. Vladykov, Vadim D. 1943. Relation between fish and fish-eating birds. Can. Field-Nat. 57 (7/8): 124–132, 0 fig. (Contr. Dépt. Pêch. Québec (12)).

122. *Vladykov, V.D. 1944. Recherches sur le

Marsouin blanc ou Béluga. Rapp. Gén. Dept. Pêch. mar., Québec (1944) pp. 102–105, ? fig.

123. Vladykov, Vadim-D. [sic]. 1944. Études sur les mammifères aquatiques. III.- Chasse, biologie et valeur économique du Marsouin blanc ou Béluga (*Delphinapterus leucas*) du fleuve et du golfe Saint-Laurent. Contr. Dépt. Pêch. Québec (14): 1–194, 57 fig. [*see* Vladykov, V.D. 1981, translation into English].

124. Vladykov, V.-D. [sic]. 1945. Rapport du biologiste du Départment des Pêcheries. Rapp. Gén. Dépt. Pêch. mar., Québec (1944) pp. 43–54, 0 fig. (Contr. Dépt. Pêch. Québec (16)).

125. Vladykov, Vadim D. 1945. Trois poissons nouveaux pour la Province de Québec. Naturaliste can. 72(1/2): 27–39, 5 fig. (Contr. Dépt. Pêch. Québec (15)).

126. Vladykov, Vadim-D. [sic]. 1946. Études sur les mammifères aquatiques. IV.- Nourriture du Marsouin blanc ou Béluga (*Delphinapterus leucas*) du fleuve Saint-Laurent. Contr. Dépt. Pêch. Québec (17): 1–155, 42 fig. [*see* Vladykov, V.D. 1981, translation into English].

127. Vladykov, Vadim D. & Gérard Beaulieu. 1946. Études sur l'esturgeon (*Acipenser*) de la Province de Québec. I-Distinction entre deux espèces d'Esturgeon par le nombre de boucliers osseux et de branchiospines. Naturaliste can. 73(6/8): 143–204, 23 fig. (Contr. Dépt. Pêch. Québec (18)).

128. Vladykov, V.-D. [sic]. 1946. Rapport du biologiste. Département des Pêcheries. Rapp. Gén. Dépt. Pêch., Québec (1945–46) pp. 46–54, 3 fig. (Contr. Dépt. Pêch. Québec (20)).

129. Vladykov, Vadim D. 1947. Nouveau bar (*Lepibema chrysops*) pour la Province de Québec. Naturaliste can. 74(7/8): 195–206, 2 fig. (Contr. Dépt. Pêch. Québec (21)).

130. Vladykov, V.-D. [sic]. 1947. Rapport du biologiste du Département des Pêcheries. Rapp. Gén. Dépt. Pêch., Québec (1946–47) pp. 44–61, 10 fig. (Contr. Dépt. Pêch. Québec (22)).

131. Vladykov, Vadim-D. [sic]. 1947. Taxonomic characters and growth of the metamorphosed northern lamprey (*Ichthyomyzon unicuspis*) from Quebec. Copeia (4): 291 (title only).

132. Vladykov, Vadim-D. [sic]. 1947. Comparison between exterior characters of two species of sturgeon from Quebec. Copeia (4): 291 (title only).

133. Vladykov, V.-D. [sic]. 1948. Rapport du biologiste du Département des Pêcheries. Rapp. Gén. Dépt. Pêch., Québec (1947–48) pp. 50–57, 2 fig. (Contr. Dépt. Pêch. Québec (25) [sic]).

134. Vladykov, Vadim D. 1948. Movements of Quebec shad as demonstrated by tagging. Canadian Fisherman 35(11): 10–11, 3 fig.

135. Vladykov, Vadim D. & Jean-Marie Roy. 1948. Biologie de la lamproie d'eau douce (*Ichthyomyzon unicuspis*) après la métamorphose. (Résumé). Rev. Can. Biol. 7(3): 483–485, 0 fig. (Contr. Dépt. Pêch. Québec (25)).

136. Vladykov, Vadim D. 1949. Quebec lampreys (Petromyzonidae) I. – List of species and their economical importance. Contr. Dépt. Pêch. Québec (26): 1–67, 21 fig.

137. Vladykov, V.-D. [sic]. 1949. Rapport du biologiste du Département des Pêcheries. Rapp. Gén. Dépt. Pêch., Québec (1948–49) pp. 50–67, 7 fig. (Contr. Dépt. Pêch. Québec (27)).

138. Vladykov, V.D. 1949. Captures de poissons d'eau salée en eau douce et de poissons d'eau douce en eau salée dans le Québec. Annales de l'ACFAS 15: 85, 0 fig.

139. Vladykov, V.D. 1949. Les Cottidés d'eau douce du Québec. Annales de l'ACFAS 15: 85, 0 fig.

140. Vladykov, V.D. 1949. Le saumon de l'Atlantique (Salmo salar) aux environs de Québec. Annales de l'ACFAS 15: 86, 0 fig.

141. Vladykov, V.D. 1949. Liste préliminaire des lamproies du Québec. Annales de l'ACFAS 15: 86, 0 fig.

142. Vladykov, V.D. 1949. Description of ammocoetes of three Quebec lampreys. Copeia (4): 306 (title only).

143. Vladykov, Vadim D. 1950. Larvae of eastern American lampreys (Petromyzonidae) I.- Species with two dorsal fins. Naturaliste can.

77(3/4): 73–95, 13 fig. (Contr. Dépt. Pêch. Québec (29)).

144. Vladykov, Vadim D. 1950. Movements of Quebec shad (*Alosa sapidissima*) as demonstrated by tagging. Naturaliste can. 77(5/6): 121–135, 4 fig. (Contr. Dépt. Pêch. Québec (30)).

145. Vladykov, V.-D. [sic]. 1950. Rapport du biologiste du Département des Pêcheries. Rapp. Gén. Dépt. Pêch., Québec (1949–50) pp. 54–76, 5 fig. (Contr. Dépt. Pêch. Québec (31)).

146. Vladykov, Vadim D. 1951. The capture of the hagfish (*Myxine glutinosa*) in the Gulf of St. Lawrence, Quebec. Copeia (1): 84, 0 fig. (Contr. Dépt. Pêch. Québec (32)).

147. Vladykov, Vadim D. 1951. Fecundity of Quebec lampreys. Can. Fish Cult. (10): 1–14, 1 fig. (Contr. Dépt. Pêch. Québec (33)).

148. Vladykov, V.D. 1951. Rapport du biologiste du Département des Pêcheries. Rapp. Gén. Dépt. Pêch., Québec (1950–51) pp. 54–71, 5 fig. (Contr. Dépt. Pêch. Québec (35)).

149. Vladykov, Vadim D. 1951. Freshwater fisheries in the Arctic. 112 pp., 0 fig. [Treats Nearctic and Palearctic fisheries] (manuscript).

150. Vladykov, Vadim D. & Gérard Beaulieu. 1951. Études sur l'esturgeon (*Acipenser*) de la Province de Québec. II. Variation du nombre de branchiospines sur le premier arc branchial. Naturaliste can. 78: 129–154, 6 fig. (Contr. Dépt. Pêch. Québec (34)).

151. Vladykov, Vadim D. 1952. Distribution des lamproies (Petromyzonidae) dans la Province de Québec. Naturaliste can. 79: 85–120, 13 fig. (Contr. Dépt. Pêch. Québec (36)).

152. Vladykov, Vadim D. 1952. Qu'est-ce qu'une sardine? Carnets Soc. Zool. de Québec 12(2): 57–62, 3 fig. (Contr. Dépt. Pêch. Québec (37)).

153. Vladykov, Vadim D. 1952. Présence dans le Québec du *Morone americana*, troisième espèce des serranidés. Naturaliste can. 79(12): 325–330, 1 fig. (Contr. Dépt. Pêch. Québec (38)).

154. Vladykov, Vadim D. 1952. Rapport du Laboratoire de Limnologie. Rapp. Gén. Dépt. Pêch., Québec (1951–52) pp. 60–77, 7 fig. (Contr. Dépt. Pêch. Québec (39)).

155. Vladykov, Vadim D. & David H. Wallace. 1952. Studies of the striped bass, *Roccus saxatilis* (Walbaum), with special reference to the Chesapeake Bay region during 1936–1938. Bull. Bingham Oceanogr. Coll. 14(1): 132–177, 7 fig.

156. Vladykov, Vadim D. 1953. Anatomical distinctions between three eastern North American species of sturgeon (Acipenser). Copeia (4): 250 (title only).

157. Vladykov, Vadim D. 1953. Tentative taxonomy of ammocoetes of the genus Ichthyomyzon. Copeia (4): 250 (title only).

158. Vladykov, V.D. 1953. Rapport du Laboratoire de Limnologie. Rapp. Gén. Dépt. Pêch. Québec (1952–53) pp. 60–88, 6 fig. (Contr. Dépt. Pêch. Québec (41)).

159. Vladykov, Vadim D. 1953. Truite mouchetée / Speckled trout. Département des Pêcheries, Province de Québec, Poissons du Québec, Album 1: 1–8, 2 pl.

160. Vladykov, Vadim D. 1953. Saumon de l'Atlantique. L'Ami du Pêcheur 1953: 66–67, 0 fig.

161. Vladykov, Vadim D. 1953. Saumon de l'Atlantique / Atlantic salmon. Département des Pêcheries, Province de Québec, Poissons du Québec, Album 2: 1–8, 2 pl.

162. Vladykov, Vadim D. 1953. Homard d'Amérique / American lobster. Département des Pêcheries, Province de Québec, Poissons du Québec, Album 3: 1–8, 2 pl.

163. Vladykov, Vadim D. & Bernard W. Corson. 1953. The taxonomic position of the Sunapee trout (*Salvelinus aureolus*). Proc. Northeast Div. Amer. Fish. Soc., Sept. 21–23, 1953, New Hampshire Fish Game Dept. 9 pp., 0 fig.

164. Homans, R.E.S. & V.D. Vladykov. 1954. Relation between feeding and the sexual cycle of the haddock. J. Fish. Res. Board Can. 11(5): 535–542, 3 fig. (Contr. Dépt. Pêch. Québec (44)).

165. Vladykov, Vadim D. 1954. Taxonomic characters of the eastern North America [sic] chars (*Salvelinus* and *Cristivomer*). J. Fish. Res. Board Can. 11(6): 904–932, 12 fig. (Contr. Dépt. Pêch. Québec (42)).

166. Vladykov, Vadim D. 1955. *Lampetra zanandreai,* a new species of lamprey from northern Italy. Copeia (3): 215–223, 4 pl. (Contr. Dépt. Pêch. Québec (45)). [published 19 August 1955; *Lampetra zanandreai* n. sp.].
167. Vladykov, Vadim D. 1955. Rapport du Laboratoire de Limnologie. Rapp. Gén. Dépt. Pêch., Québec (1953–54) pp. 40–60, 2 fig. (Contr. Dépt. Pêch. Québec (46)).
168. Vladykov, Vadim D. 1955. A comparison of Atlantic sea sturgeon with a new subspecies from the Gulf of Mexico (*Acipenser oxyrhynchus de sotoi*). J. Fish. Res. Board Can. 12(5): 754–761, 8 pl. (Contr. Dépt. Pêch. Québec (47)). [published September 1955; *Acipenser oxyrhynchus de sotoi* n. subsp.].
169. Vladykov, Vadim D. 1955. Les morues / Cods. Département des Pêcheries, Province de Québec, Poissons du Québec, Album 4: 1–12, 2 pl.
170. Vladykov, Vadim D. 1955. Les esturgeons / Sturgeons. Département des Pêcheries, Province de Québec, Poissons du Québec, Album 5: 1–11, 3 fig., 2 pl.
171. Vladykov, Vadim D. 1955. L'Anguille / Eel. Départment des Pêcheries, Province de Québec, Poissons du Québec, Album 6: 1–12, 2 pl.
172. Vladykov, Vadim D. 1955. Remarks on the anatomy of western American sturgeons. Copeia (4): 319 (title only).
173. Vladykov, Vadim D. 1955. Pacific species of lampreys of the genus Entosphenus. Copeia (4): 319 (title only).
174. Vladykov, Vadim D. 1956. Fecundity of wild speckled trout (*Salvelinus fontinalis*) in Quebec Lakes. J. Fish. Res. Board Can. 13(6): 799–841, 10 fig. (Contr. Dépt. Pêch. Québec (51)).
175. Vladykov, Vadim D. 1956. Remarks on the osteology of American Salmonidae. Copeia (4): 272 (title only).
176. Vladykov, Vadim D. 1956. Distant recaptures of shad (*Alosa sapidissima*) tagged in Québec. Naturaliste can. 83(10): 235–248, 4 fig. (Contr. Dépt. Pêch. Québec (54)).
177. Vladykov, Vadim D. 1957. Fish tags and tagging in Quebec waters. Trans. Amer. Fish. Soc. 86(1956): 345–349, 0 fig. (Contr. Dépt. Pêch. Québec (55)).
178. Vladykov, Vadim D. 1957. Les formes locales de la truite rouge du Québec (*Salvelinus marstoni*). Naturaliste can. 84(12): 233–248, 0 fig. (Contr. Dépt. Pêch. Québec (56)) [published December 1957; *Salvelinus marstoni intermedius* n. subsp., *Salvelinus marstoni cavanaghi* n. subsp.].
179. Vladykov, Vadim V. 1957. Échouement de grosses baleines près de Québec. Actualités marines, Québec 1(3): 22–25, 4 fig.
180. Vladykov, Vadim D. 1957. The present trend in American ichthyology: progress or retrogress? 8 pp., 0 fig. (manuscript).
181. Vladykov, Vadim D. & Adolphe Michaud. 1957. Les voyages de l'éperlan. Actualités marines, Québec 1(1): 15–19, 3 fig.
182. Vladykov, V.D. 1958. Deux pêches caractéristiques du fleuve Saint-Laurent. Actualités marines, Québec 2(1): 6–12, 4 fig.
183. Vladykov, Vadim D. 1958. Lampetra fluviatilis Linné, lamproie européenne, vit-elle sur le continent nord-américain? Annales de l'ACFAS 24: 74, 0 fig.
184. Vladykov, Vadim D. 1958. Lampetra planeri (Bloch), lamproie européenne, vit-elle sur le continent Nord-américain [sic]? Annales de l'ACFAS 24: 75, 0 fig.
185. Vladykov, Vadim D. 1958. Les espèces de lamproies nord-américaines de (sic) genres Lampetra et Entosphenus devraient être classées dans le seul genre Lampetra? Annales de l'ACFAS 24: 75, 0 fig.
186. Vladykov, Vadim D. 1958. Baleine à bosse (Megaptera novaeangliae) échouée à Lauzon en Septembre 1957. Annales de l'ACFAS 24: 80, 0 fig.
187. Vladykov, V.D. 1958. Méthode nouvelle d'étiquetage des poissons. Annales de l'ACFAS 24: 80, 0 fig.
188. Vladykov, Vadim D. & G. Beaulieu. 1958. Variations dans le Québec de l'abondance annuelle des poissons originaires des Grands-Lacs. Naturaliste can. 85(6/7): 149–156, 3 fig. (Contr. Dépt. Pêch. Québec (70)).

189. Vladykov, Vadim D. & Gérard Beaulieu. 1958. Variation dans l'abondance annuelle à Québec des poissons originaires des Grands Lacs. Annales de l'ACFAS 24: 78, 0 fig.
190. Vladykov, Vadim D. & Jacques Brousseau. 1958. Croissance du Bar d'Amérique, Roccus saxatilis, dans le Québec. Annales de l'ACFAS 24: 79, 0 fig.
191. Vladykov, Vadim D. & W.I. Follett. 1958. Redescription of *Lampetra ayresii* (Günther) of western North America, a species of lamprey (Petromyzontidae) distinct from *Lampetra fluviatilis* (Linnaeus) of Europe. J. Fish. Res. Board Can. 15(1): 47–77, 15 fig. (Contr. Dépt. Pêch. Québec (68)).
192. Vladykov, Vadim D. & Jean-Marie Roy. 1958. Est-ce qu'il y a une distinction entre les différentes populations de lamproie marine, P. marinus Linné? Annales de l'ACFAS 24: 75, 0 fig.
193. Vladykov, Vadim D. 1959. Review of: 'Freshwater Fishery Biology' by Karl F. Lagler, Wm. C. Brown Company, Publishers, Dubuque, Iowa, 1956; 434 pp., 214 figures. Cloth bound $6.75. Can. Fish. Cult. (23) (1958): 49, 0 fig.
194. Vladykov, V.-D. [sic]. 1959. Notes sur la biologie de l'anguille d'Amérique (*Anguilla rostrata*). Annales de l'ACFAS 25: 93–94, 0 fig.
195. Vladykov, Vadim D. 1959. The taxonomy of European ammocoetes (Petromyzontidae). (Abstract). XVth Inter. Congress of Zoology, London, Section II, Paper 38: 207–208, 0 fig.
196. Vladykov, Vadim D. 1959. The true taxonomic position of the lake trout (*Cristivomer namaycush*). Trans. Northeast Wildlife Conference, Tenth Annual Meeting, Montréal, Que., Canada, January 4, 5, 6, 7, 1958: 74, 0 fig.
197. Vladykov, Vadim D. 1959. The effects on fisheries of man-made changes in fresh water in the Province of Quebec. Can. Fish. Culturist. 25: 7–12, 0 fig.
198. Vladykov, Vadim D. 1959. Local populations of the marine lamprey (*Petromyzon marinus*). 7 pp., 0 fig. (manuscript).
199. Clark, John R. & Vadim D. Vladykov. 1960. Definition of haddock stocks of the northwestern Atlantic. U.S. Fish and Wildlife Service, Fish. Bull. 169, vol. 60: 283–296, 7 fig.
200. Vladykov, Vadim D. 1960. Description of young ammocoetes belonging to two species of lampreys: *Petromyzon marinus* and *Entosphenus lamottenii*. J. Fish. Res. Board Can. 17(2): 267–288, 13 pl.
201. Vladykov, Vadim D. 1960. Geographical variation in the size of eel elvers, *Anguilla rostrata*. Copeia (4): 402 (title only).
202. Vladykov, Vadim D. 1960. Quelques observations sur l'anguille électrique (Electrophorus electricus) en captivité. Annales de l'ACFAS 26: 95, 0 fig.
203. Vladykov, Vadin-V. [sic]. 1961. Variations géographiques dans la taille des civelles de l'anguille d'Amérique (*Anguilla rostrata*). (Title only). Annales de l'ACFAS 27: 66, 0 fig.
204. Vladykov, V.D. 1961. Do species change? 26 pp., 0 fig. [lecture, apparently given 4 April 1961]. (manuscript).
205. Vladykov, Vadim D. [& D.E. McAllister]. 1961. Preliminary list of marine fishes of Quebec. Naturaliste can. 88(3): 53–78, 0 fig. [published March 1961; *Melanostigma normani* Vladykov nomen nudum].
206. Vladykov, V.D. & G.N. Mukerji. 1961. Order of succession of different types of infraoral lamina in landlocked sea lamprey (*Petromyzon marinus*). J. Fish. Res. Board Can. 18(6): 1125–1143, 7 fig.
207. Vladykov, Vadim D. 1962. Osteological studies on Pacific salmon of the genus *Oncorhynchus*. Fish. Res. Board Can. Bulletin 136: 1–172, 89 fig.
208. Vladykov, Vadim D. 1963. A review of salmonid genera and their broad geographical distribution. Trans. Roy. Soc. Can. Ser. IV, Sec. III, 1: 459–504, 17 fig.
209. Vladykov, Vadim D. & John R. Greeley. 1963. *Order* Acipenseroidei. pp. 24–60, fig. 7–10. *In:* Fishes of the Western North Atlantic. Memoir Sears Found. Mar. Res. I, Part 3: 1–630 (Contr. Dépt. Pêch. Québec (53)).

210. Vladykov, Vadim D. 1964. Report to the Government of Iran on the inland fisheries resources of Iran, especially of the Caspian Sea with special reference to sturgeon. Rep. FAO / Exp. Prog. Tech. Ass., Rome (1818): 1–51, 0 fig.
211. Vladykov, Vadim D. 1964. Review of: Wonders of wildlife in Europe by Franz Roedelberger and Vera Groschoff. (Longmans Canada, Toronto. 232 pp. Translation). Can. Geog. Journal 69(1): VII, 0 fig.
212. Vladykov, Vadim D. 1964. Quest for the true breeding area of the American eel (*Anguilla rostrata* LeSueur). J. Fish. Res. Board Can. 21(6): 1523–1530, 3 fig.
213. Vladykov, Vadim D. 1964. The relative abundance of the adult males and females of the eel of America (*Anguilla rostrata*) in the salt and the soft waters of the West Atlantic coast. Annales de l'ACFAS 30: 65, 0 fig.
214. Vladikov, Vadem D. [sic]. 1965. Où l'aire de ponte de l'anguille d'Amérique, *Anguilla Rostrata*, se situe-t-elle? Annales de l'ACFAS 31: 54, 0 fig.
215. Vladykov, V.D. 1965. Genera of the parasitic lampreys (Petromyzontidae) from the Pacific region and their broad geographical distribution. 43 pp., 0 fig. (manuscript).
216. Vladykov, V.D. 1965. Intraspecific variability of sea lamprey, *Petromyzon marinus* Linnaeus. 84 pp., 0 fig. (manuscript).
217. Vladykov, Vadim D. & W.I. Follett. 1965. *Lampetra richardsoni,* a new nonparasitic species of lamprey (Petromyzonidae) from western North America. J. Fish. Res. Board Can. 22(1): 139–158, 9 fig. [published January 1965; *Lampetra richardsoni* n. sp.].
218. Vladykov, Vadim D. 1966. Remarks on the American eel (Anguilla rostrata LeSueur). Sizes of elvers entering streams; the relative abundance of adult males and females; and present economic importance of eels in North America. Verh. Internat. Verein. Limnol. 16: 1007–1017, 3 fig.
219. Vladykov, Vadim D. 1966. Apodal leptocephali of the western Atlantic. Copeia (4): 907 (title only).
220. James, D.T. & V.D. Vladykow [sic]. 1967. A method of preparing eel otoliths for age determination. Annales de l'ACFAS 33: 69, 0 fig.
221. Vladykov, Vadim D. 1967. Morphological pecularities of the apodal leptocephali. Annales de l'ACFAS 33: 70–71, 0 fig.
221. Vladykov, Vadim D. 1967. Progress report no. 1 of the American eel (*Anguilla rostrata*) studies in Canada. Indust. Dev. Branch, Fish. Serv., Dept. Fish. and Forestry, Ottawa, Progress Report (1): 1–6, 0 fig.
223. Vladykov, Vadim D. 1967. Elvers of the American eel (Anguilla rostrata) in the Maritime Provinces. Indust. Dev. Branch, Fish. Serv., Dept. Fish. and Forestry, Ottawa, Progress Report (2): 7–31, 4 fig.
224. Vladykov, V.D. 1967. Age determination and age of the American eel (*Anguilla rostrata*) from New Brunswick waters. Indust. Dev. Branch, Fish. Serv., Dept. Fish. and Forestry, Ottawa, Progress Report (3): 32–52, 1 fig.
225. Vladykov, Vadim D. & W.I. Follett. 1967. The teeth of lampreys (Petromyzonidae): Their terminology and use in a key to the Holarctic genera. J. Fish. Res. Board Can. 24(5): 1067–1075, 4 fig.
226. Beaulieu, Gérard & Vadim D. Vladykov. 1968. Comportement de l'anguille (Anguilla rostrata) du fleuve St-Laurent selon les résultats d'étiquetage effectués de 1945 à 1957. Annales de l'ACFAS 35: 95, 0 fig.
227. Gruchy, C.G. & V.D. Vladykov. 1968. Sexual dimorphism in anal fin of brown trout, *Salmo trutta,* and close relatives. J. Fish. Res. Board Can 25(4): 813–815, 3 fig.
228. Vladykov, V.D. 1968. Lengths of American eel (*Anguilla rostrata*) in commercial catches of eastern Canadian provinces. Indust. Dev. Branch, Fish. Serv., Dept. Fish. and Forestry, Ottawa, Progress Report (4): 53–73, 1 fig.
229. Vladykov, V.D. 1968. Eel tagging in New Brunswick waters. I. Length and colouration of tagged eels. Indust. Dev. Branch, Fish. Serv., Dept. Fish. and Forestry, Ottawa, Progress Report (5): 74–98, 5 fig.
230. Vladykov, Vadim D. 1968. Study of eel pop-

ulations in Crecy Lake, headwater tributary of the Bocabec River, New Brunswick. Indust. Dev. Branch, Fish. Serv., Dept. Fish. and Forestry, Ottawa, Progress Report (6): 1–26, 3 fig.

231. Vladykov, Vadim D. & Donald T. James. 1968. Sex composition of eels (*Anguilla rostrata*) in Crecy Lake, New Brunswick. Annales de l'ACFAS 35: 93, 0 fig.

232. * Vladykov, V.D. 1969. Eels. pp. 196–199, ? fig. *In:* The Encyclopedia of Marine Resources, Van Nostrand Reinhold Company, New York.

233. Vladykov, Vadim D. 1969. A study of the growth and development of elvers of the American eel (*Anguilla rostrata*) transported from New Brunswick to a freshwater pond in Ontario. Indust. Dev. Branch, Fish. Serv., Dept. Fish. and Forestry, Ottawa, Progress Report (7): 1–38, 1 fig.

234. Vladykov, Vadim D. 1969. Eel tagging in Cape Breton, Nova Scotia, 1968. Indust. Dev. Branch, Fish. Serv., Dept. Fish. and Forestry, Ottawa, Progress Report (8): 1–29, 0 fig.

235. Vladykov, Vadim D. 1969. Investigation on the causes of mortality of two shipments of live eels from Moncton to Europe by Air Canada in 1968. Indust. Dev. Branch, Fish. Serv., Dept. Fish. and Forestry, Ottawa, Progress Report (9): 1–43, 5 fig.

236. Vladykov, Vadim D. & Richard H. Backus. 1969. Distribution and growth of *Anguilla* Leptocephali in the western North Atlantic, based on collections by Woods Hole research vessels, 1953–1967. Copeia (4): 875 (title only).

237. Vladykov, Vadim D. 1970. Pearl tubercles and certain cranial peculiarities useful in the taxonomy of coregonid genera. pp. 167–193, 24 fig. *In:* C.C. Lindsey & C.S. Woods (ed.) Biology of Coregonid Fishes, University of Manitoba Press, Winnipeg.

238. Vladykov, Vadim D. 1970. Homing of the American eel (*Anguilla rostrata*). Indust. Dev. Branch, Fish. Serv., Dept. Fish. and Forestry, Ottawa, Progress Report (10): 1–26, 2 fig.

239. Vladykov, Vadim D. 1970. Methods for keeping eels alive during air transport from Canada to Europe. Indust. Dev. Branch, Fish. Serv., Dept. Fish. and Forestry, Ottawa, Progress Report (11): 1–34, 22 fig.

240. Vladykov, Vadim D. 1970. Project for the development of commercial eel fisheries in insular Newfoundland. Indust. Dev. Branch, Fish. Serv., Dept. Fish. and Forestry, Ottawa, Progress Report (12): 1–23, 2 fig.

241. Vladykov, Vadim D. 1971. Homing of the American eel, *Anguilla rostrata,* as evidenced by returns of transplanted tagged eels in New Brunswick. Can. Field-Nat. 85(3): 241–248, 1 fig.

242. Vladykov, Vadim D. 1972. Sous-division en trois sous-familles des lamproies de l'hémisphère-nord de la famille Petromyzonidae. Annales de l'ACFAS 39: 148, 0 fig.

243. Vladykov, Vadim D. 1972. Lampreys of the Ottawa area. Trail & Landscape, Ottawa 6(4): 121–128, 4 fig.

244. Vladykov, V.D. 1972. Morphological differences in male gonads among nine genera of Gadidae (Pisces). J. Fish. Res. Board Can. 29(12): 1709–1716, 8 fig.

245. Vladykov, V.D. & C.G. Gruchy. 1972. Comments on the nomenclature of some subgenera of Salmonidae. J. Fish Res. Board Can. 29(11): 1631–1632, 0 fig. [published 6 November 1972; *Parahucho* n. subg., *Parasalmo* n. subg.].

246. Vladykov, Vadim D. & Sylvie Pharand. 1972. Tentacules du velum ('velar tentacles') chez les lamproies de l'Amérique du Nord. Annales de l'ACFAS 39: 148, 0 fig.

247. Vladykov, V.D. 1973. Petromyzonidae. pp. 2–5, 0 fig. *In:* J.C. Hureau & Th. Monod (ed.) Check-list of the Fishes of the North-Eastern Atlantic and of the Mediterranean (CLOFNAM). Vol. I. UNESCO, Paris.

248. Vladykov, V.D. 1973. Myxinidae. p. 6, 0 fig. *In:* J.C. Hureau & Th. Monod (ed.) Check-list of the Fishes of the North-Eastern Atlantic and of the Mediterranean (CLOFNAM). Vol. I. UNESCO, Paris.

249. Vladykov, Vadim D. 1973. A female sea lam-

prey (*Petromyzon marinus*) with a true anal fin, and the question of the presence of an anal fin in Petromyzonidae. Can. J. Zool. 51(2): 221–224, 4 fig.

250. Vladykov, V.D. 1973. *Lampetra pacifica,* a new nonparasitic species of lamprey (Petromyzontidae) from Oregon and California. J. Fish. Res. Board Can. 30(2): 205–213, 9 fig. [published 6 February 1973; *Lampetra pacifica* n. sp.].
251. Vladykov, Vadim D. 1973. Macrophthalmia in the American eel (*Anguilla rostrata*). J. Fish. Res. Board Can. 30(5): 689–693, 2 fig.
252. Vladykov, Vadim D. 1973. North American nonparasitic lampreys of the family Petromyzonidae must be protected. Can. Field-Nat. 87(3): 235–239, 0 fig.
253. Vladykov, V.D. 1974. Request for a ruling on the stem of the family-group names based on the type-genus *Petromyzon* Linnaeus, 1758. Z.N. (S.) 2045. Bull. zool. Nomencl. 30(3/4): 198–199, 0 fig.
254. Vladykov, Vadim D., Herratt March & Peter Liew. 1974. Duration of the leptocephalous stage of *Anguilla rostrata* (Le Sueur). Copeia (4): 1021 (title only).
255. Vladykov, V.D. & C.G. Gruchy. 1975. Request for a ruling on the stem of the family-group name based on the type-genus *Petromyzon* Linnaeus, 1758: Revised application. Z.N.(S.) 2045. Bull. zool. Nomencl. 32(3): 154–155, 0 fig.
256. Vladykov, V.D., Peter Liew & R.E. Whitfield. 1975. Size and age of the American eel (*Anguilla rostrata*) collected in the fishway at the Moses-Saunders Dam, Cornwall, Ontario. 11 pp., 4 fig., 5 tabs. (manuscript).
257. Vladykov, Vadim D. & Herratt March. 1975. Distribution of leptocephali of the two species of *Anguilla* in the western North Atlantic based on collections made between 1933 and 1968. Syllogeus 6: 1–38, 4 fig.
258. Vladykov, Vadim D. & Edward E. Kott. 1976. The taxonomic significance of velar tentacles in holarctic lampreys (Petromyzonidae). 2ième Congrès Européen des Ichtyologistes, Paris, Septembre 1976, Abstract no. 111, 6 pp. 0 fig.
259. Vladykov, Vadim D. & Edward Kott. 1976. Is *Okkelbergia* Creaser and Hubbs, 1922 (Petromyzonidae) a distinct taxon? Can. J. Zool. 54(3): 421–425, 2 fig.
260. Vladykov, Vadim D. & Edward Kott. 1976. A second nonparasitic species of *Entosphenus* Gill, 1862 (Petromyzonidae) from Klamath River system, California. Can. J. Zool. 54(6): 974–989, 8 fig. [published 18 June 1976; *Entosphenus folletti* n. sp.].
261. Vladykov, V.D. & E.E. Kott. 1976. The taxonomic significance of velar tentacles in Holarctic lampreys (Petromyzonidae). Rev. Trav. Inst. Pêches marit. 40(3/4): 787–789, 0 fig.
262. Vladykov, V.D. & Edward Kott. 1976. Preliminary list of Holarctic lampreys (Petromyzonidae) and their broad distribution. Wilfrid Laurier Univ., Museum Zool., Waterloo, Contr. 1: ii + 1–5, 0 fig.
263. Vladykov, Vadim D. & Edward Kott. 1976. A new nonparasitic species of lamprey of the genus *Entosphenus* Gill, 1862, (Petromyzonidae) from south central California. Bull. South. Cal. Acad. Sci. 75(2): 60–67, 3 fig. [published August 1976; *Entosphenus hubbsi* n. sp.].
264. Vladykov, V.D., Edward Kott & S. Pharand-Coad. 1976. A new nonparasitic species of lamprey, genus *Lethenteron* (Petromyzonidae), from eastern tributaries of the Gulf of Mexico, U.S.A. Publ. Zool., Natl. Mus. Nat. Sci., Natl. Mus. Can., Ottawa (12): 1–36, 6 fig. [published 16 January 1976; *Lethenteron meridionale* n. sp.].
265. Gruchy, Charles G. & Vadim D. Vladykov. 1977. Viscera and the systematics of *Acipenser*. Copeia (4): 821 (title only).
266. Vladykov, Vadim D. & Edward Kott. 1977. Satellite species among the Holarctic lampreys (Petromyzonidae). (Abstract no. 621). Amer. Zool. 17(4), 0 fig.
267. Vladykov, Vadim D. & Edward Kott. 1978. A new nonparasitic species of the Holarctic lamprey genus *Lethenteron* Creaser and Hubbs, 1922, (Petromyzonidae) from northwestern North America with notes on other

species of the same genus. Biol. Pap. Univ. Alaska 19: 1–74, 13 fig. [published April 1978; *Lethenteron alaskense* n. sp.].

268. Vladykov, Vadim D. & Edward Kott. 1978. A new nonparasitic species of *Lethenteron* Creaser and Hubbs 1922 (Petromyzonidae) from Hokkaido, Japan. Can. J. Zool. 56(8): 1792–1800, 6 fig. [published 25 August 1978; *Lethenteron matsubarai* n. sp.].

269. Vladykov, Vadim D. & Edward Kott. 1979. A new parasitic species of the holarctic lamprey genus *Entosphenus* Gill, 1862 (Petromyzonidae) from Klamath River, in California and Oregon. Can. J. Zool. 57(4): 808–823, 8 fig. [published 10 May 1979; *Entosphenus similis* n. sp.].

270. Vladykov, Vadim D. & Edward Kott. 1979. Satellite species among the holarctic lampreys (Petromyzonidae). Can. J. Zool. 57(4): 860–867, 8 fig.

271. Vladykov, Vadim D. & Edward Kott. 1979. Erratum: A new nonparasitic species of *Lethenteron* Creaser and Hubbs 1922 (Petromyzonidae) from Hokkaido, Japan. Can. J. Zool. 57(4): 951, 0 fig.

272. Vladykov, Vadim D. & Edward Kott. 1979. List of Northern Hemisphere lampreys (Petromyzonidae) and their distribution / Liste des lamproies (Petromyzonidae) de l'hémisphère nord et leur distribution. Misc. Spec. Pub. Dept. Fish. Oceans, Ottawa (42): 1–30, 12 fig.

273. Vladykov, Vadim D. & Edward Kott. 1980. Description and key to metamorphosed specimens and ammocoetes of Petromyzonidae found in the Great Lakes region. Can. J. Fish. Aquat. Sci. 37(11): 1616–1625, 2 fig.

274. Vladykov, Vadim D. & Edward Kott. 1980. First record of the sea lamprey, *Petromyzon marinus L.*, in the Gulf of Mexico. Northeast Gulf Science 4(1): 49–50, 0 fig.

275. Vladykov, V.D., E. Kott & D.E. McAllister. 1980. *Lampetra alaskanse* [sic] (Vladykov and Kott), Darktail lamprey. p. 22, 2 fig. *In:* D.S. Lee et al. (ed.) Atlas of North American Freshwater Fishes, North Carolina State Museum of Natural History, Raleigh.

276. Vladykov, V.D. 1981. Studies of aquatic mammals III. Hunting, biology and economic value of the white whale or beluga (*Delphinapterus leucas*) of the St. Lawrence River and Gulf. Can. Translation Fish. Aquat. Sci. 4739: 1–222, 57 fig. [English translation of Vladykov, V.D. 1944. Études sur les mammifères aquatiques III. . . .].

277. Vladykov, V.D. 1981. Studies of aquatic mammals IV. Food of the white whale or beluga (*Delphinapterus leucas*) of the St. Lawrence river. Can. Translation Fish. Aquat. Sci. 4723: 1–184, 42 fig. [English translation of Vladykov, V.D. 1946. Études sur les mammifères aquatiques. IV. . . .].

278. Vladykov, Vadim D. & Edward Kott. 1982. Correct scientific names for the least brook lamprey and the American brook lamprey (Petromyzontidae). Can. J. Zool. 60(5): 856–864, 2 fig.

279. Vladykov, Vadim D. & Edward Kott. 1982. Comment on Reeve M. Bailey's view of lamprey systematics. Can. J. Fish. Aquat. Sci. 39(8): 1215–1217, 0 fig.

280. Vladykov, V.D. & P.K.L. Liew. 1982. Sex of adult American eels (*Anguilla rostrata*) collected as elvers in two different streams along the eastern shore of Canada, and raised in the same freshwater pond in Ontario. Proceedings of the 1980 North American Eel Conference, Ontario Ministry of Natural Resources. Ont. Fish. Tech. Rep. Ser. 4: 88–93, 0 fig.

281. Vladykov, Vadim D., Claude B. Renaud, Edward Kott & Panos S. Economidis. 1982. A new nonparasitic species of Holarctic lamprey, genus *Eudontomyzon* Regan 1911 (Petromyzontidae), from Greece. Can. J. Zool. 60(11): 2897–2915, 10 fig. [published 10 December 1982; *Eudontomyzon hellenicus* n. sp.].

282. Fernholm, Bo & V.D. Vladykov. 1984. Myxinidae. pp. 68–69, 7 fig. *In:* P.J.P. Whitehead, M.-L. Bauchot, J.-C. Hureau, J. Nielsen & E. Tortonese (ed.) Fishes of the North-eastern Atlantic and the Mediterranean Vol. 1, UNESCO, Paris.

283. Vladykov, V.D. 1984. Petromyzonidae. pp.

64–67, 9 fig. *In:* P.J.P. Whitehead, M.-L. Bauchot, J.-C. Hureau, J. Nielsen & E. Tortonese (ed.) Fishes of the North-eastern Atlantic and the Mediterranean Vol. 1, UNESCO, Paris.

284. Vladykov, Vadim D. & Edward Kott. 1984. A second record for California and additional morphological information on *Entosphenus hubbsi* Vladykov and Kott 1976 (Petromyzontidae). California Fish and Game 70(2): 121–127, 2 fig.
285. Vladykov, Vadim D. 1985. Does neoteny occur in Holarctic lampreys (Petromyzontidae)? Syllogeus 57: 1–13, 2 fig.
286. Vladykov, V.D. 1985. Record of 61 parasitic lampreys (*Ichthyomyzon unicuspis*) on a single sturgeon (*Acipenser fulvescens*) netted in the St. Lawrence River (Québec). Naturaliste can. 112(3): 435–436, 2 fig.
287. Vladykov, Vadim D. 1985. Hagfish. p. 784. 0 fig. *In:* The Canadian Encyclopedia, Hurtig Publishers, Edmonton. Volume II, p. 669–1370.
288. Vladykov, Vadim D. 1985. Lamprey. p. 970. 0 fig. *In:* The Canadian Encyclopedia, Hurtig Publishers, Edmonton. Volume II, p. 669–1370.
289. Vladykov, Vadim D., Claude B. Renaud & Sylvie Laframboise. 1985. Breeding tubercles in three species of *Gadus* (cods). Can. J. Fish. Aquat. Sci. 42(3): 608–615, 4 fig.
290. Kott, Edward, Claude B. Renaud & Vadim D. Vladykov. 1988. Urogenital papillae in the Holarctic lampreys (Petromyzontidae). Env. Biol. Fish. (this issue).

Acknowledgements

We are grateful to M. Vianney Legendre for his detailed comments on a rough draft of this bibliography. Edward Kott reviewed an early draft of the manuscript.

Environmental Biology of Fishes Vol. 23, No. 1–2, pp. 37–43, 1988

The urogenital papilla in the Holarctic lamprey (Petromyzontidae)

Edward Kott[1], Claude B. Renaud [2] & Vadim D. Vladykov[3]
[1] *Department of Biology, Wilfrid Laurier University, Waterloo, Ontario N2L 3C5, Canada*
[2] *Ichthyology Section, National Museum of Natural Sciences, Ottawa, Ontario K1P 6P4, Canada*
[3] *Department of Biology, University of Ottawa, Ottawa, Ontario K1N 6N5, Canada*

Received 23.12.1987 Accepted 5.1.1988

Key words: Taxonomy, Parasitic species, Nonparasitic species, Sexual dimorphism, Variation

Synopsis

The urogenital papillae of 30 species of Holarctic lampreys were examined to determine if this structure has any taxonomic use. Total length, branchial length and papilla length were measured. A correlation between mean papilla length and mean total length of the species existed in males for parasitic species but not nonparasitic species, indicating a potential use in nonparasitic species. In nonparasitic species the most obvious differences existed in the papilla length: branchial length ratio. From the limited material available there appeared to be no geographic variation in this ratio, although there is a seasonal variation with the ratio increasing as spawning approaches. It therefore appears that urogenital papilla length has taxonomic value in male nonparasitic lampreys, provided the specimens are collected in spawning condition. The female papilla is too small to be of use.

Introduction

Lampreys spawn only once during their lifetime and die shortly thereafter. As the spawning period approaches, the secondary sexual characteristics become expressed. In males, the urogenital papilla becomes elongate, the cloacal lips swell, the tail bends down sharply and the second dorsal fin becomes heightened, serrated, and vascularized at the base (Vladykov 1949, Smith et al. 1968, Hardisty & Potter 1971). In females, the papilla elongates slightly, an anal fin-like fold or keel develops posterior to the cloaca, the region in front of the cloaca swells, the tail turns sharply upwards, the height of the second dorsal fin increases while it becomes serrated and vascularized, and a swelling develops on the leading edge (Vladykov 1949, Smith et al. 1968, Hardisty & Potter 1971).

The development of some of these characters is under the control of pituitary hormones. Hypophysectomy in autumn and winter inhibits the development of secondary sexual characteristics in both sexes. In spring, it causes growth of the papilla to stop and regression of the height of the dorsal fin and cloacal swelling in males; however, in females, the anal fin-like fold continues to develop for a time (Larsen & Rothwell 1972).

Vladykov & Kott (1982 a, b) have suggested that one of these features, the length of the urogenital papilla, may be of taxonomic significance. This has been proven useful in the genus *Eudontomyzon* (Vladykov et al. 1982). In order to investigate the importance of this character, we studied 30 species of Holarctic lampreys.

This papilla has had a varied terminology. It has been described as a genital papilla (Vladykov 1949), a penis (Hardisty & Potter 1971), a urinogenital papilla (Hardisty 1979) and a urogenital papilla (Smith et al. 1968, Vladykov & Kott 1982 a, b). The first two terms are inappropriate for they imply that this papilla has only a reproductive function. Since this structure also has a urinary function, we recommend that it be called a urogenital papilla.

Materials and methods

Only the papillae of sexually mature individuals were examined for species comparisons. Other than to illustrate a sexual dimorphism in length, most of the studies involved only males.

The urogenital papilla was measured from the point of its appearance out of the cloacal slit to its extremity. Although part of the papilla is internal, only its exposed part was measured to avoid time-consuming and destructive dissections. Furthermore, the papilla cannot be retracted or protracted, and thus its exposed part is indicative of the lamprey's state of maturity.

Total length (TL) and branchial length (B_1–B_7) were measured for each specimen as described by Vladykov & Follett (1965). In addition, the following ratios were compared: papilla length/TL, papilla length/B_1–B_7, and B_1–B_7/TL.

Results

The various species studied, selected body proportions, and papilla length are summarized in Tables 1–3. Differences exist between species in the relative length of the papilla with the longest papillae found in the genera *Lampetra, Lethenteron,* and *Eudontomyzon.* The species with the longest papilla relative to total length or branchial length was

Table 1. Species of Holarctic lampreys with length of urogenital papilla in spawning males 20–40% of branchial length (B_1–B_7). Figures in brackets represent ranges.

Species	N	TL mm	B_1–B_7 mm	papilla mm	$\frac{\text{papilla}}{B_1-B_7}$ %	$\frac{\text{papilla}}{\text{TL}}$ %	$\frac{B_1-B_7}{\text{TL}}$ %
Lampetra aepyptera	70	118.8 (102–146)	12.8 (10.0–16.0)	5.2 (3.0–7.5)	40.8 (23.1–60.0)	4.4 (2.2–5.8)	10.8 (9.1–13.1)
Lethenteron meridionale	1	112	13.0	5.0	38.5	4.5	11.6
Lethenteron reissneri	3	147 (143–154)	16.0 (15.0–17.0)	6.0 (4.0–7.0)	37.2 (26.7–48.8)	4.2 (2.6–5.0)	10.9 (9.7–11.9)
Lampetra planeri	10	124.8 (111–140)	12.5 (9.5–15.0)	4.4 (3.0–6.0)	36.2 (21.4–52.6)	3.6 (2.1–4.6)	9.9 (8.5–11.5)
Lethenteron zanandreai	6	134.7 (128–139)	15.4 (14.0–17.0)	4.8 (4.0–6.0)	31.0 (23.5–40.0)	3.5 (2.9–4.5)	11.4 (10.6–12.3)
Eudontomyzon mariae	8	171.1 (143.5–184)	19.9 (16.0–22.0)	5.9 (4.0–8.5)	29.7 (25.0–38.6)	3.5 (2.8–4.6)	11.6 (10.7–12.3)
Ichthyomyzon gagei	6	102.7 (92–111)	10.3 (9.0–11.0)	2.8 (2.5–3.0)	27.6 (22.7–33.3)	2.8 (2.3–3.3)	10.1 (9.2–10.8)
Lethenteron alaskense	1	141	13	3.5	26.9	2.5	9.2
Lampetra richardsoni	26	134.3 (122–151)	13.7 (12.0–16.0)	3.4 (2.5–4.0)	24.9 (17.6–30.8)	2.5 (1.7–3.1)	10.2 (9.3–11.5)
Lampetra pacifica	10	125.9 (112–133)	13.2 (11.0–15.0)	3.1 (2.5–4.5)	23.8 (17.8–36.4)	2.5 (2.0–3.6)	10.5 (9.8–11.3)
Lampetra fluviatilis	19	255 (197–283)	25.5 (17.0–29.5)	6.0 (3.0–9.0)	23.3 (15.9–37.5)	2.3 (1.5–3.7)	10.0 (8.6–10.6)
Ichthyomyzon greeleyi	23	130.5 (108–156)	13.3 (10.0–17.0)	2.8 (2.0–4.0)	21.3 (16.0–28.6)	2.2 (1.5–2.7)	10.2 (8.5–11.7)

Table 2. Species of Holarctic lampreys with length of urogenital papilla in spawning males 10–19% of branchial length (B_1–B_7). Figures in brackets represent ranges.

Species	N	TL mm	B_1–B_7 mm	papilla mm	papilla / B_1–B_7 %	papilla / TL %	B_1–B_7 / TL %
Caspiomyzon wagneri	5	327.8 (301–344)	34.1 (26.0–38.0)	5.8 (5.0–7.0)	17.3 (14.3–21.2)	1.8 (1.5–2.2)	10.3 (8.6–11.6)
Lethenteron japonicum	6	377.2 (339–401)	40.0 (38.0–42.0)	6.8 (6.0–8.0)	17.1 (14.6–19.5)	1.8 (1.5–2.2)	10.7 (9.5–11.4)
Eudontomyzon danfordi	1	180.0	18.5	3.0	16.2	1.7	10.3
Entosphenus lethophagus	1	129	12.5	2.0	16.0	1.6	9.7
Entosphenus minimus	12	81 (78–100)	8.8 (7.0–10.0)	1.1 (1.0–1.5)	12.4 (10.0–18.8)	1.3 (1.0–1.8)	10.1 (9.0–11.6)
Entosphenus tridentatus	2	214 (211–217)	24.5 (24.0–25.0)	3.0	12.3 (12.0–12.5)	1.4	11.5 (11.4–11.5)
Entosphenus hubbsi	9	113.7 (106–124)	12.9 (12.0–14.0)	1.6 (1.0–2.0)	12.3 (8.3–16.7)	1.4 (0.9–1.8)	11.3 (10.5–12.4)
Ichthyomyzon bdellium	10	217.5 (117–247)	23.6 (21.0–26.0)	3.0 (2.0–4.0)	12.3 (8.7–16.0)	1.3 (0.9–1.8)	11.2 (9.7–18.8)
Lampetra ayresii	2	180.5 (174–187)	19.8 (19.5–20.0)	2.3 (2.0–2.5)	11.4 (10.3–12.5)	1.3 (1.2–1.3)	11.0 (10.7–11.2)
Petromyzon marinus	7	435.3 (385–492)	45.7 (39.0–52.0)	5.1 (4.0–6.0)	11.2 (9.5–13.0)	1.2 (0.9–1.4)	10.5 (9.9–11.6)
Ichthyomyzon fossor	25	119.5 (99–149.5)	13.5 (10.5–16.0)	1.4 (1.0–2.5)	10.5 (6.7–21.7)	1.2 (0.7–2.4)	10.8 (9.3–12.0)

Table 3. Species of Holarctic lampreys with length of urogenital papilla in spawning males less than 10% of branchial length (B_1–B_7). Figures in brackets represent ranges.

Species	N	TL mm	B_1–B_7 mm	papilla mm	papilla / B_1–B_7 %	papilla / TL %	B_1–B_7 / TL %
Tetrapleurodon geminis	4	119.8 (115–126)	11.8 (11.0–13.0)	1.1 (1.0–1.5)	9.6 (7.7–12.5)	1.0 (0.8–1.3)	9.8 (9.2–10.4)
Lethenteron wilderi	44	169.1 (146–214)	19.5 (15.0–25.0)	1.9 (1.0–3.0)	9.4 (4.5–16.7)	1.1 (0.7–1.7)	11.5 (10.5–12.2)
Lethenteron matsubarai	3	158.3 (147–169)	15.7 (15.0–17.0)	1.2 (1.0–1.5)	7.5 (5.9–10.0)	0.7 (0.6–1.0)	9.9 (9.4–10.2)
Ichthyomyzon unicuspis	7	236.3 (185–279)	25.8 (23.5–28.5)	1.7 (1.0–2.5)	6.8 (5.3–7.7)	0.8 (0.4–1.1)	11.0 (10.2–12.7)
Eudontomyzon hellenicus	8	110.3 (96–133)	12.6 (11.0–15.0)	0.8 (0.5–1.0)	6.5 (3.3–9.1)	0.8 (0.4–1.0)	11.5 (10.7–12.5)
Entosphenus folletti	1	210	17.5	1.0	5.7	0.5	8.3
Ichthyomyzon castaneus	1	183	19.0	1.0	5.3	0.5	10.4

Lampetra aepyptera, a nonparasitic form. Of those species with a papilla length 20–40% of the branchial length (Table 1), only one was parasitic (*Lampetra fluviatilis*). This represents 8.3% of the species in this category. For those with a papilla 10–19% of branchial length, seven of ten or 70% were parasitic (Table 2), with *C. wagneri* not included since its feeding habits are not definitely known (Holčík 1986). Only two of seven species or 28.6% were parasitic in the group for which the papilla was less than 10% of branchial length (Table 3). It appears that nonparasitic species tend to have a very long or short papilla whereas parasitic species have one of intermediate length.

If one considers the cases in which stem species occur sympatrically with their satellite species (Vladykov & Kott 1979), a pattern emerges (Table 4). Where such pairs occur, the nonparasitic form has a relatively longer papilla. Where data were available, the following sources of variation in relative length were examined: sexual dimorphism, time of collection, site of collection, and body length.

The urogenital papilla was shorter in females (Table 5). In *Petromyzon marinus,* the male papilla was about twice as long whereas in *Lampetra aepyptera,* it was almost five times as long. A difference in shape was also noted: males have a cone-shaped papilla while in females, it is trough-shaped.

Only for *L. aepyptera* were sufficient specimens available; these were collected at different times of the year from the same geographic location (Table 6). Average papilla length/B_1–B_7 for males collect-

Table 4. Comparison of selected stem and satellite species of Holarctic lampreys with respect to male urogenital papilla length.

Parasitic stem species	Mean papilla length as % of branchial length	Nonparasitic satellite species	Mean papilla length as % of branchial length
I. castaneus	5.3	*I. gagei*	27.6
I. unicuspis	6.8	*I. fossor*	10.5
I. bdellium	12.3	*I. greeleyi*	21.3
L. ayresii	11.4	*L. richardsoni*	24.9
L. japonicum	17.1	*L. reissneri*	37.2
L. fluviatilis	23.3	*L. planeri*	36.2

Table 5. Comparison of urogenital papilla length in males and females of selected species of Holarctic lampreys. Figures in brackets represent ranges.

Species	Males				Females			
	TL mm	papilla mm	papilla length / B_1–B_7 %	N	TL mm	papilla mm	papilla length / B_1–B_7 %	N
L. aepyptera	118.8 (102–146)	5.2 (3.0–7.5)	40.8 (23.1–60.0)	70	120.4 (105–140)	1.1 (1.0–1.5)	9.2 (6.7–17.5)	7
L. richardsoni	134.3 (122–151)	3.4 (2.5–4.0)	24.9 (17.9–30.8)	26	101.5 (100–103)	1.0	10.6 (10.0–11.1)	2
L. fluviatilis	255 (197–283)	6.0 (3.0–9.0)	23.3 (15.9–37.5)	19	255 (235–288)	2.1 (1.0–3.0)	8.2 (4.2–13.3)	11
E. tridentatus	214 (211–217)	3.0	12.3 (12.0–12.5)	2	203.2 (199–207.5)	1.2 (1.0–1.5)	5.2 (4.4–6.0)	2
P. marinus	435.3 (385–492)	5.1 (4.0–6.0)	11.2 (9.5–13.0)	7	462.1 (410–519)	2.8 (2.0–3.0)	5.9 (3.8–7.7)	9

ed on February 6, 1980 was 22.6%. By March 18, 1980 this percentage had reached 38.8% indicating that the length increases as the individuals become sexually mature.

For three species, data were available to test if a difference existed in the papilla length of mature males from different geographical areas (Table 7). No significant difference was found between the localities.

For the 30 species, the relationship between mean papilla length and mean total length of a species was investigated and weak correlation was discovered (Table 8). The data were then separated into parasitic and nonparasitic species. For non-

Table 6. Length of urogenital papilla in males of *Lampetra aepyptera* collected on two different dates from the Tombigbee River, Alabama. Figures in brackets represent ranges.

Date	TL mm	B_1–B_7 mm	papilla mm	$\frac{\text{papilla}}{B_1\text{–}B_7}$ %	N
Feb. 6, 1980	117.0 (103–134)	12.0 (10–14)	2.5 (1.5–5.0)	22.6 (11.5–41.7)	15
Mar. 18, 1980	129.0 (118–142)	14.1 (13–16)	5.4 (4.5–7.0)	38.8 (32.1–46.2)	11

Table 7. Comparison of urogenital papilla length in mature males of selected species of Holarctic lampreys, collected from different localities. Figures in brackets represent ranges.

Species	Locality	$\frac{\text{papilla}}{B_1\text{–}B_7}$ %	N
L. planeri	Spain	36.0 (29.2–45.5)	14
	Czechoslovakia	36.2 (21.4–52.6)	10
I. greeleyi	Pennsylvania	24.4 (18.7–28.6)	4
	Ohio	21.6 (19.2–25.0)	6
L. richardsoni	Washington	26.1 (21.4–30.8)	12
	British Columbia	23.9 (17.6–30.8)	14

Table 8. Summary of linear regression analysis of relation between mean papilla length and mean total length for 30 species of Holarctic lampreys.

Case	No. of points	slope	standard error	Probability slope = 0	r
All species combined	30	0.0101	0.00377	0.012	0.453
nonparasitic species only	19	– 0.00262	0.0160	>0.5	– 0.040
parasitic species only	10	0.0155	0.00437	0.007	0.781

parasitic species, no correlation existed; however, a strong positive correlation was found for parasitic species ($r = 0.781$). The weak correlation when all species were combined was due to the correlation found in parasitic species.

Finally, a very strong correlation was found ($r = 0.98$) when mean branchial length was compared to mean total length for the 30 species.

Discussion

A considerable variation occurs in the length of the urogenital papilla in males of Holarctic species of lampreys, which suggests that this character may be of taxonomic value. Although this is the first study of the character using an extensive series of species, it has been used for taxonomic purposes by Vladykov & Kott (1982 a, b) and Vladykov et al. (1982). In a study, Smith et al. (1968) found that the papilla of *Ichthyomyzon unicuspis* was shorter than the papilla of *I. fossor* and *Lethenteron wilderi*. Results similar to those of Smith et al. were found in the present study.

The emphasis has been on the male papilla since the female papilla is considerably shorter and therefore the differences between the species are not as distinct (Table 5). Also, since the papilla lengthens as lampreys approach maturity, it is important that only mature male specimens be used when measuring this character.

From the limited data available, it appears that there is little variation in the relative length of the papilla in populations from different geographic localities. This increases the value of this feature as a taxonomic character.

A comparison of different parasitic species revealed a positive correlation between the mean length of the papilla and mean species total length. Differences in papilla length for parasitic species may largely reflect length difference between the species and therefore have little taxonomic significance. In nonparasitic species, however, no such correlation existed. The lack of correlation may, at first, appear to be due to the fact that the mean size of nonparasitic species fall within a small range which is not great enough to reveal a correlation. This does not appear to be the case as significant differences occur between species of similar total length. For example, in males of *Entosphenus hubbsi* and *I. gagei* having mean lengths of 113.7 mm and 102.7 mm respectively a highly significant difference exists between their papilla length ($t = 8.98$, 13d.f., $p = 1.2 \times 10^{-5}$). This indicates that papilla length could be a valuable character for the identification of nonparasitic species. Vladykov & Kott (1982 b) stressed this when discussing the proper name for the nonparasitic American brook lamprey.

In general, the ratio B_1–B_7/TL shows little variation within a species or between species in a genus (Vladykov & Kott 1978). In this study, a strong correlation was also found when comparing this ratio between all 30 species investigated. The high correlation ($r = 0.989$) indicates that branchial length is very representative of total length. Therefore, the ratio papilla length/B_1–B_7 accurately reflects the ratio papilla length/TL. Since branchial length is shorter than total length, the use of the ratio papilla length/B_1–B_7 makes the differences between species more obvious.

In summary, it appears that male urogenital papilla length, especially when it is compared to branchial length, is an important character for describing the differences between the nonparasitic species of lampreys, providing that only sexually mature individuals are used.

This study also revealed the great difference in relative papilla length of males in sympatrically-occurring species pairs of the type that have been called stem and satellite species by Vladykov & Kott (1979). In such cases, the nonparasitic species had the relatively longer papillae. This difference may play some role in minimizing possible hybridization between such species.

References cited

Hardisty, M.W. 1979. Biology of the cyclostomes. Chapman and Hall, London. 428 pp.

Hardisty, M.W. & I.C. Potter. 1971. The general biology of adult lampreys. pp. 127–206. *In:* M.W. Hardisty & I.C. Potter (ed.) The Biology of Lampreys, Vol. 1, Academic Press, London.

Holčík, J. 1986. *Caspiomyzon wagneri*. pp. 119–142. *In:* J. Holčík (ed.) The Freshwater Fishes of Europe, Vol. 1, Part I, AULA-Verlag, Wiesbaden.

Larsen, L.O. & B. Rothwell. 1972. Adenohypophysis. pp. 1–67. *In:* M.W. Hardisty & I.C. Potter (ed.) The Biology of Lampreys, Vol. 2, Academic Press, London.

Smith, A.J., J.H. Howell & G.W. Piavis. 1968. Comparative embryology of five species of lampreys of the upper Great Lakes. Copeia 1968: 461–469.

Vladykov, V.D. 1949. Quebec lampreys. 1. List of species and their economic importance. Contr. Dép. Pêch. Québec 26: 1–67.

Vladykov, V.D. & W.I. Follett. 1965. *Lampetra richardsoni,* a new nonparasitic species of lamprey (Petromyzonidae) from western North America. J. Fish. Res. Board Can. 22: 139–158.

Vladykov, V.D. & E. Kott. 1978. A new nonparasitic species of the holarctic lamprey genus *Lethenteron* Creaser and Hubbs, 1922, (Petromyzonidae) from northwestern North America with notes on other species of the same genus. Biol. Pap. Univ. Alaska 19: 1–74.

Vladykov, V.D. & E. Kott. 1979. Satellite species among the holarctic lampreys (Petromyzonidae). Can. J. Zool. 57: 860–867.

Vladykov, V.D. & E. Kott. 1982a. Comment on Reeve M. Bailey's view of lamprey systematics. Can. J. Fish. Aquat. Sci. 39: 1215–1217.

Vladykov, V.D. & E. Kott. 1982b. Correct scientific names for the least brook lamprey and the American brook lamprey (Petromyzontidae). Can. J. Zool. 60: 856–864.

Vladykov, V.D., C.B. Renaud, E. Kott & P.S. Economidis. 1982. A new nonparasitic species of Holarctic lamprey, genus *Eudontomyzon* Regan 1911 (Petromyzontidae), from Greece. Can. J. Zool. 60: 2897–2915.

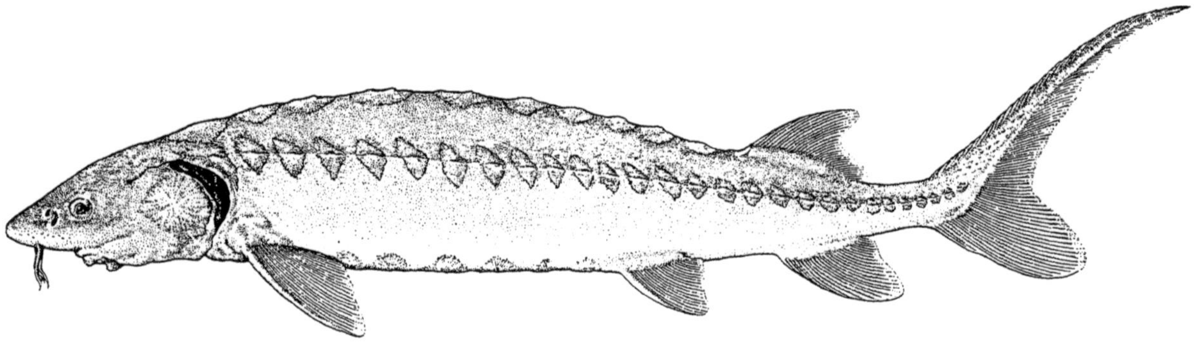

Lateral view of spawning female *Acipenser brevirostris* by Paul I. Voevodine for Vladykov & Greeley (1963).

Environmental Biology of Fishes Vol. 23, No. 1–2, pp. 45–54, 1988

Metamorphosis of the mountain brook lamprey *Ichthyomyzon greeleyi*

F. William, H. Beamish & Trudy E. Medland
Department of Zoology, Group for the Advancement of Fish Studies, University of Guelph, Guelph, Ontario N1G 2W1, Canada

Received 15.3.1987 Accepted 25.12.1987

Key words: Adult, Ammocoetes, Growth, Morphology, Ontogeny, Spawning, Cyclostomata, Petromyzontiformes

Synopsis

Metamorphosis of the nonparasitic mountain brook lamprey, *I. greeleyi*, is described within 7 phases based on external morphological changes. The commencement of metamorphosis in early August is consistent with the pattern for northern hemisphere lampreys which display an inverse relationship with latitude. The sequence of external changes including the development of eyes, oral disc, teeth, tongue, oral fimbriae and lateral line is consistent among individuals of *I. greeleyi* and lampreys in general. Some of the changes realized during metamorphosis of *I. greeleyi* are important to the adult period while others reflect their presumed parasitic ancestry. Energy reserves by nonfeeding *I. greeleyi* during the 100–140 days required to complete metamorphosis are replaced by water so that total length and body weight do not change significantly during this time.

Introduction

Characteristic of all lamprey life cycles is a distinct larval or ammocoete and adult period. A juvenile period is present in parasitic lampreys but absent in nonparasitic or brook lampreys. Following the larval period, lampreys undergo a process of metamorphosis manifested as a change in form, the result of rapid developmental processes taking place within the organism (Potter 1980, Youson 1980, Beamish & Austin 1985). These changes equip parasitic species for life as a predator of fish. Nonparasitic lampreys do not feed following metamorphosis.

In the general ecological context, parasitic and nonparasitic lampreys separate into two categories, altricial and precocial (Balon 1980, 1981, 1983) in the sense that the latter are more completely developed at a particular time in their ontogeny than the former (Beamish & Thomas 1984). Aside from some postlarval differences between the two forms, the somatic changes experienced by the nonparasitic lampreys are similar to those undergone by the parasitic forms. This process referred to as metamorphosis in precocial lamprey species is not strictly identical to that experienced by the altricial forms.

The nonparasitic mountain brook lamprey, *Ichthyomyzon greeleyi* Hubbs and Trautman, 1937, is believed to be derived from the parasitic *Ichthyomyzon bdellium* (Jordan, 1885) and occurs within many of the tributaries of the Ohio and Tennessee River systems (Vladykov & Kott 1979). It is the opinion of Vladykov & Kott (1979) that *I. greeleyi* is not distinguishable from *Ichthyomyzon hubbsi* Raney, 1952 which should then be considered a junior subjective synonym of *I. greeleyi*.

The present study describes the process of metamorphosis for the mountain brook lamprey within 7 phases (Youson & Potter 1979, Beamish & Thomas 1984). Metamorphosis was considered complete when the external changes, particularly those in the head and branchial regions, assumed the appearance of an adult.

Materials and methods

Ammocoetes, metamorphosing and adult *I. greeleyi* were collected by electrofishing from four tributaries of the French Broad River in western North Carolina (U.S.A.), Bent Creek, Cane Creek, Mills River and Davidson River between 1980 and 1986. Some metamorphosing animals were transported live to the University of Guelph where they were maintained in tanks supplied with a continuous flow of nonchlorinated ground water. A suitable substrate for burrowing was supplied (15 cm) and water temperatures were adjusted regularly to correspond to those of their natal streams (Beamish & Austin 1985). Other samples of lamprey were killed by an overdose of methane tricaine sulfonate (Sandoz) and preserved in 5% formalin or frozen in water for subsequent examination. Preserved animals were measured in the laboratory for total length and weight (± 1 mm and ± 0.001 g). The relationships between live and preserved total lengths and body weights are described for *I. greeleyi* by the regressions (± 95% C.L.):

$$L = 1.06 \pm 0.08\ P_l - 3.53 \pm 0.24$$
$$(n = 22:\ r = 0.980,\ P<0.05),$$

where L is live length and P_l is preserved total length (both in mm) and

$$W = 0.87 \pm 0.17\ P_w - 0.21 \pm 0.20$$
$$(n = 22;\ r \pm 0.940,\ P<0.05),$$

where W is live weight and P_w is preserved weight (both in mg).

Sex was determined by examination of the gonads under a dissection microscope. Histological examination of gonads was periodically conducted confirming the identifications made by visual inspection. Ovaries of some animals were weighed (± 10 mg).

The four streams examined varied appreciably in size. Bent Creek (Buncombe County) was the shortest with an approximate length of 6 km between the mouth and a dam, above which is located a large reservoir. Davidson River (Transylvania County) is about 17 km in length with one major tributary. Mills River (Henderson County) has three major tributaries with the distance between the mouth and the source of the longest tributary, South Fork, being about 30 km. Cane Creek (Henderson and Buncombe Counties) is roughly 24 km in length and has several tributaries. The most comprehensive collection of metamorphosing and adult *I. greeleyi* was made in Bent Creek. Relatively few animals were captured in Mills River.

Total hardness and total alkalinity, 30 ± 13 (± 95% C.L.) and 28 ± 3 $mg.L^{-1}$ (as $CaCO_3$) respectively, were similar among the four streams, while the pH varied between 7.0–7.5. Temperature of Bent Creek was highest, 19–20°C, in mid August and decreased to 9 and 3°C in early October and early February, respectively. Between early February and early May, temperature increased to 14°C. Temperatures of Davidson and Mills River were within one degree of that for Bent Creek. Cane Creek was approximately 2°C warmer than Bent Creek for comparable seasons.

Results

External diagnostic changes which characterize each of the 7 phases of metamorphosis are described in Table 1 and illustrated in Figures 1 and 2. Comparative information is provided for ammocoete, adult (phase 8) and senescent (phase 9) periods. Coloration of live mountain brook lamprey did not change between the ammocoete and senescent periods and ranged from butterscotch to olive brown. The dorsal surface was generally darker than the lateral and ventral surfaces. Scattered dark pigment spots are present along the length of the dorsal-lateral surface. Occasionally, an ammocoete was found which exhibited very little pigmentation. These cream colored animals represented fewer than 1% of the population.

In ammocoetes, the branchial region is pouch-

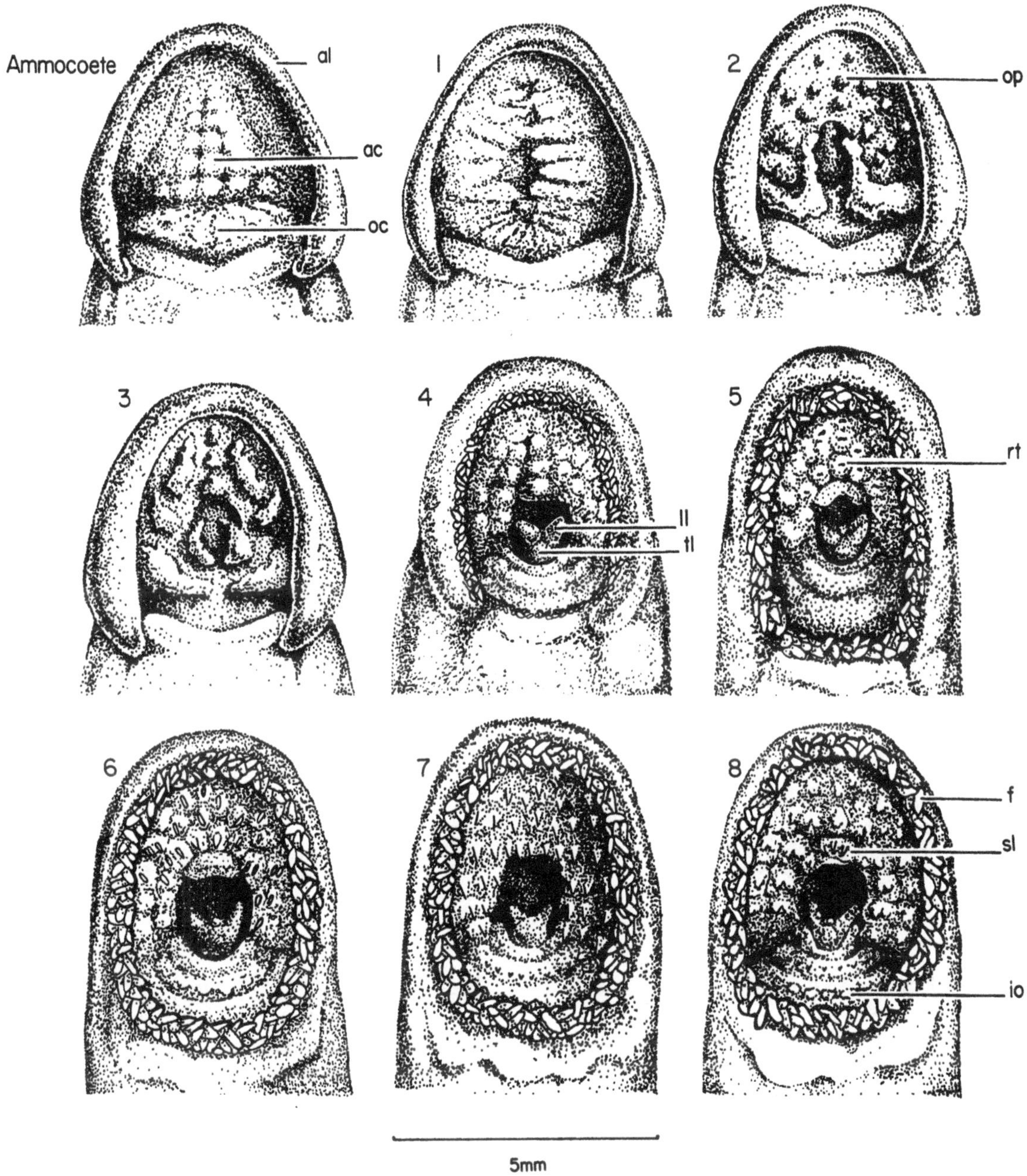

Fig. 1. Development of the oral hood and disc in metamorphosing (phases 1–7) and adult (phase 8) *I. greeleyi* (ac, anterior cirrhi; al, anterior lip; f, oral fimbriae; io, infraoral lingual lamina; ll, longitudinal lingual lamina; oc, oral cirrhi; op, oral papilla; pl, posterior transverse lip; rt, rudimentary disc teeth; so, supralingual lamina; tl, transverse lingual lamina).

like and extends below the natural contour of the ventral surface; it is bright red in color due to an abundant blood supply near the surface. By phase 4 of metamorphosis, the depth of the branchial region is reduced so that the body tapers gradually toward the tail. The reddish coloration present in ammocoetes is largely absent by phase 6. The dorsal, caudal and ventral fins are confluent; they are not well developed in ammocoetes. Maximum height of the dorsal fin in ammocoetes is about 1.7% of total length. The succession of phases through metamorphosis is accompanied by a pro-

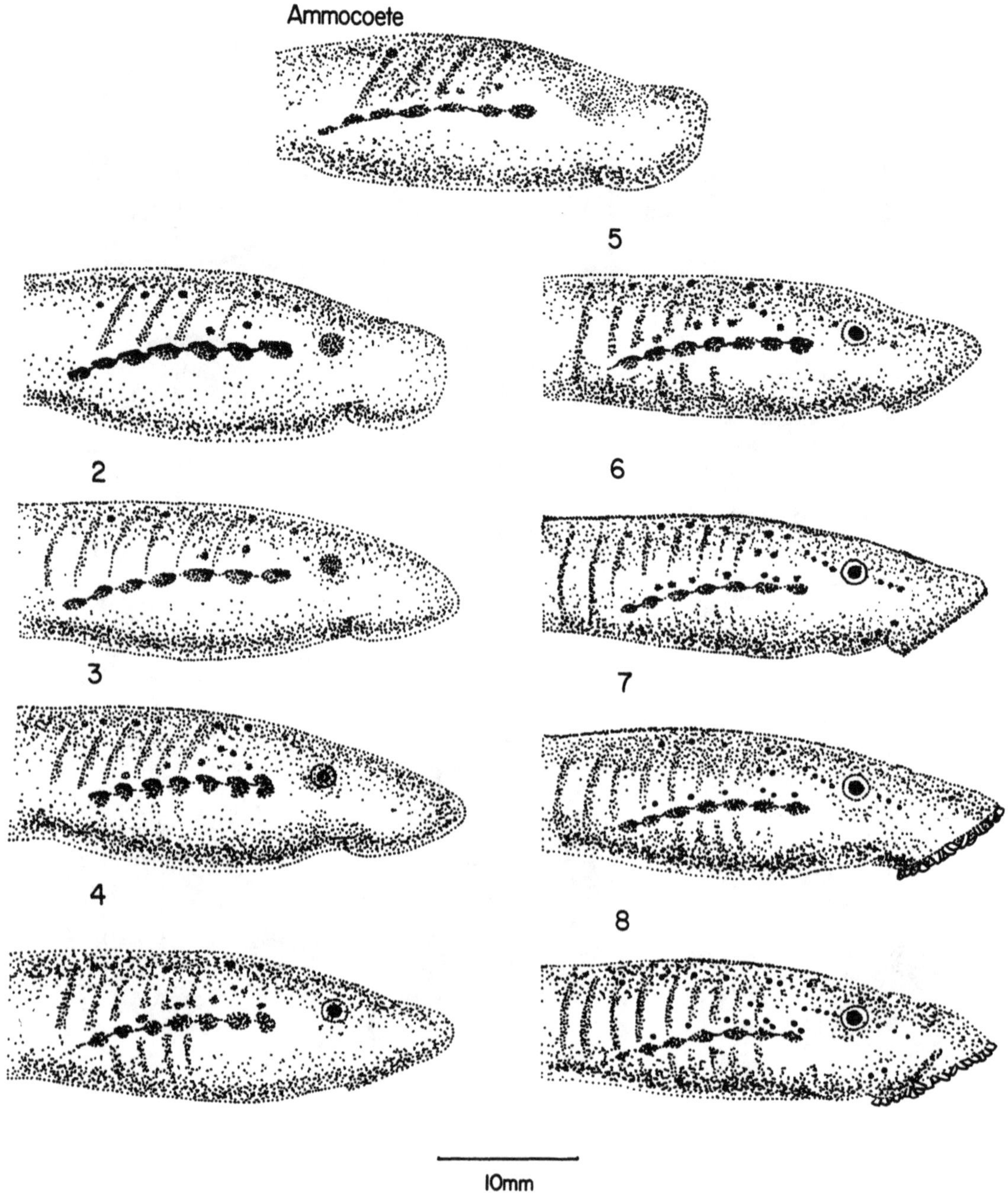

Fig. 2. Lateral view of the anterior portion of *I. greeleyi* during metamorphosis (phases 1–7) and as an ammocoete and adult (phase 8).

gressive increase in dorsal fin height which continues to the adult period. By the final phase of metamorphosis and in young adults, maximum dorsal fin height was 2.2 and 2.7% of total length, respectively. In phase 6 animals, modest emargination develops between the posterior dorsal and caudal fins.

Mountain brook lamprey in the early stages of metamorphosis were found in the sandy silt substratum along with ammocoetes. As metamorphosis progressed, preference shifted in favor of a coarser substratum and a faster water current. However, not all lampreys moved from the soft substratum during or even following metamorphosis.

It was not unusual to capture adults along with ammocoetes.

Metamorphosis commenced between early and mid August. In a large collection of animals made from Bent Creek (July 26–29, 1984), no metamorphosing animals were found. By mid August, metamorphosis was well underway. Metamorphosing mountain brook lamprey in phases 1–3 were collected from Bent Creek on August 18, 1980 (Fig. 3). In 1986, metamorphosing brook lamprey in phases 1–4 were captured between August 13–17 in Bent Creek, Cane Creek and Davidson River. By late September, metamorphosis had progressed beyond phase 4 and was completed in December.

Table 1. Diagnostic external characteristics of metamorphosis and of the ammocoete, adult and senescent periods in *I. greeleyi*.

	Phase	Oral hood/disc	Eye	General
Ammocoete		Lateral posterior ends of thin 'U' shaped anterior lip of oral hood overlap posterior transverse lip but separated by a deep notch. Prominent short anterior cirrhi and feather-like oral cirrhi.	Scarcely visible as a diffuse dark patch.	Branchiopores oval and joined by distinct longitudinal groove.
Metamorphosis	1	Oral hood opening reduced due to thickening of lips. Partial fusion between posterior ends of anterior and posterior lips. Oral cirrhi reduced.	Dark patch.	
	2	Anterior cirrhi prominent and papillae-like. Tongue (piston) sometimes visible. No indentation behind posterior lip.	Dark and enlarged with periphery well defined.	Snout narrow. Indentation on dorsal surface of snout immediately anterior to the eye ('bottle-nose').
	3	Thickening and fusion of lips provides for rectangular disc-like shape. Papillae reduced. Tongue and lingual lamina distinct but soft and without cusps.	Maximum diameter. Slight differentiation of pupil and iris.	Snout more pointed. Indentation of dorsal surface of snout less prominent.
	4	Presumptive disc oblong, lips fused into a continuous ring with lines of fusion evident. Oral fimbriae present but short. Supra-oral, infraoral and transverse lingual laminae present but not well developed.	Pupil and iris clearly differentiated.	Branchiopores oval to circular.
	5	Disc oval with fusion lines absent. Oral cirrhi and papillae reduced or absent.		
	6	Disc enlarged and more circular. White disc teeth prominent but soft and blunt. Distinct tricuspid supraoral and multicuspid infraoral laminae. Small indentation between posterior edge of oral disc and branchial region.	Projects from body contour.	Dark apertures of lateral line system conspicuous along a line posterior to disc and along the lateral and dorsal surfaces of branchial region.
	7	Teeth and laminae hardened, the former with sharp tips. Crescent shaped depression posterior to oral disc. Oral fimbriae well developed.		Branchiopores oval and longitudinal groove absent.
Adult	8	Disc teeth becoming blunt. Oral fimbriae slightly longer.		Nasopore elevated. Mature females with distended trunk and often upturned tails. In mature males urinogenital papilla conspicuous.
Senescent	9	Some teeth often missing.		Body slim lesions.

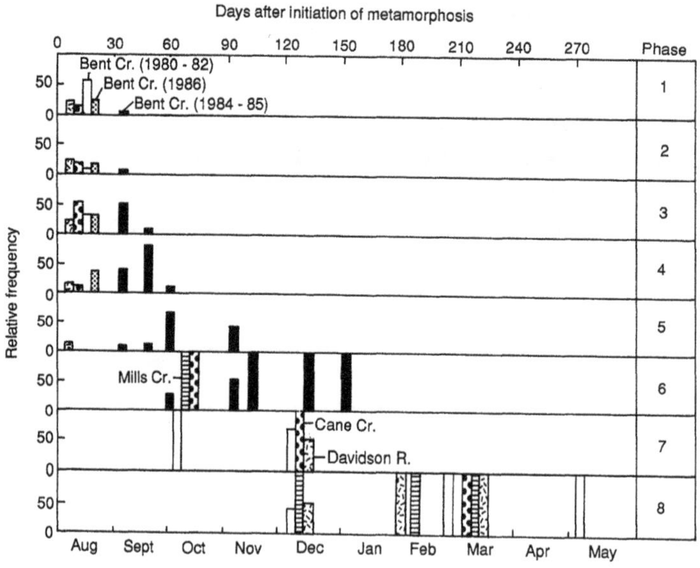

Fig. 3. Timing of metamorphosis (phases 1–7) and the adult period (phase 8) in *Ichthyomyzon greeleyi* from Bent Creek (1980–82; n = 135, 1984–85, n = 320; 1986, n = 48), Cane Creek (1980–82, n = 36, 1986, n = 50), Mills River (1980–81, n = 19) and Davidson River (1980–86, n = 39). The number of animals from each stream within each phase is presented as a percentage of the total number of metamorphosing and adult animals examined for each sampling date. Modified after Beamish & Austin (1985).

Time required for metamorphosis of the populations of mountain brook lamprey in Bent Creek, Cane Creek and Davidson River is estimated at 100–140 days.

Total live length at the commencement of metamorphosis varied from 111 to 182 mm (n = 200) in the four streams. Within a stream mean length did not change significantly throughout metamorphosis (P>0.05). Mean lengths for mountain brook lamprey in phases 1 through 7 were 130.8 ± 1.9 (n = 126), 135.0 ± 3.3 (n = 29) and 150.1 ± 2.6 mm (n = 86) in Bent Creek, Davidson River and Cane Creek, respectively. The number of animals collected from Mills River (n = 19) was insufficient to evaluate length changes. Significant differences in length were not demonstrable between Bent Creek and Davidson River at P>0.05. Metamorphosing brook lamprey from Cane Creek were significantly longer than those from either Bent Creek or Davidson River (P<0.05).

Metamorphosing brook lampreys from Bent Creek representing phases 1 to 7 were measured by sex for total length and weight. Neither length nor body weight of animals preserved in formalin changed significantly (P>0.05) between phase 1 and 7 for either males or females. Mean length and weight for males (n = 55), 132.3 ± 2.6 mm and 4420 ± 286 mg, were less but not significantly different from those for females (n ± 31), 136.3 ± 3.4 mm and 4908 ± 396 mg.

In adult mountain brook lampreys (phase 8), total length declined slightly but not significantly (P>0.05). Body weight tended to increase among females and to remain unchanged in males. Female adult *I. greeleyi* preserved in formalin were longer and heavier than adult males; however, significant differences were not demonstrable. Mean total length and weight for adult male and female brook lamprey from Bent Creek (n = 56) were 127.6 ± 2.3 and 134.6 ± 12.7 mm and 4326 ± 284 and 5431 ± 1707 mg, respectively. Similarly, total length and weight for adult males and females from Cane Creek (n = 46) were 136.3 ± 4.6 and 141.6 ± 3.4 mm and 5285 ± 590 and 5738 ± 383 mg, respectively.

Weight (mg) changes relative to length (mm) are represented by the condition factor ($W/L3 \cdot 10^3$; Ricker 1975) and illustrated for ammocoetes, metamorphosing animals and adults in Figure 4. Among large ammocoetes from Bent Creek, the

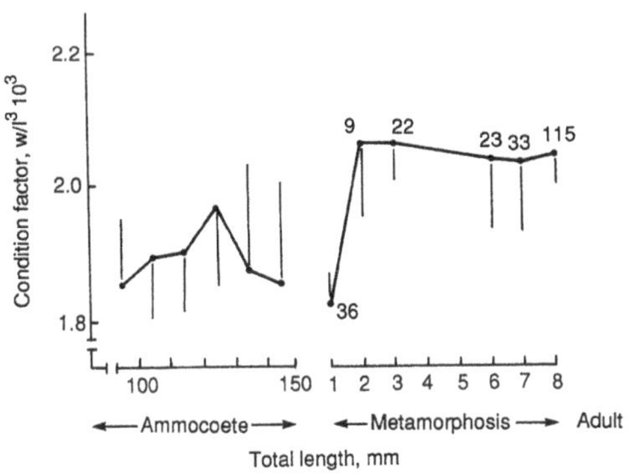

Fig. 4. Condition factor of ammocoetes, metamorphosing *I. greeleyi* (phases 1–7) and adults (phase 8) based on formalin preserved total lengths (mm) and weights (mg). Sample size was 10 for each length category of ammocoetes. The number of metamorphosing and adult *I. greeleyi* sampled is indicated beside each mean value. Vertical lines represent the 95% confidence limits.

condition factor for animals preserved in formalin did not change with length in a consistent pattern. Mean condition factor for ammocoetes of 90–149 mm was 1.888 ± 0.087 (n = 60). Between phases 1 and 2 of metamorphosis, the condition factor increased from 1.818 to 2.058 (Fig. 4). Significant changes in the condition factor were not demonstrable between phases 2 and 8 ($P<0.05$). The condition factor for males and females did not differ significantly for comparable phases of development. Further, the condition factor for metamorphosing animals did not vary significantly among the three streams from which extensive collections were made. The mean condition factor for metamorphosing (phases 2–7) and adult (phase 8) mountain brook lamprey from Bent and Cane Creeks and Davidson River was 2.046 ± 0.030 (n = 203).

Sex was determined for ammocoetes above 60 mm, the approximate length at which ovarian growth becomes marked in *I. greeleyi*. From a sample of 932 ammocoetes, the sex ratio was near parity with 1♂: 0.962♀. Sex ratio of metamorphosing brook lamprey shifted to a preponderance of males at 1♂: 0.831♀ (n = 130). The gonadosomatic ratio for large female ammocoetes (>140 mm), 0.4 ± 0.1% (n = 10) increased significantly to 0.8 ± 0.3% during phases 1–3 (n = 10) and abruptly to 12.6 ± 2.9% (n = 10) in recently metamorphosed animals. The gonadosomatic ratio in adult females approximately 2 weeks before spawning (May 5) was 21.5 ± 1.6% (n = 10).

Discussion

Commencement of metamorphosis in the northern hemisphere is regulated within the summer season by latitude and modified by environmental conditions. Lampreys from the higher latitudes in the northern hemisphere typically begin metamorphosis earlier than those from the more southern regions (Fig. 5). Few species of lampreys are found in the southern hemisphere and metamorphosis has been described only for *Geotria australis* (Potter et al. 1980).

The sequence of changes in external metamorphosis in *I. greeleyi* is remarkably consistent with that for other lamprey species, both parasitic and nonparasitic (Table 2). This is to be predicted on the basis of saltatory ontogeny (Balon 1986). Thus, structures that collectively form an organ or system pace their rates of development to become complete simultaneously and begin a new function. In lampreys, this is well illustrated in the transition of the oral hood and larval alimentary canal to an oral disc and adult alimentary canal (Hilliard et al. 1983, Beamish & Thomas 1984). Alterations in the alimentary canal and the acquisition of an oral disc accomodate changes in feeding habits between the ammocoete and juvenile periods. Nonparasitic lampreys reflecting their origin from closely related parasitic species undergo similar changes to their alimentary canal although in at least some species, development is not always complete (Battle & Hayashida 1965, Beamish & Thomas 1984).

The time required to complete metamorphosis by *I. greeleyi*, 100–140 days, is similar to that by *Lampetra japonica kessleri* (90–105 days; Poltorykhina 1971), *Lampetra tridentata* (105–120 days; Richards 1980) but it is short in comparison to some species. Metamorphosis in *G. australis*, *Ichthyomyzon gagei* and *Lampetra planeri* requires 150–160, 170–210 and 210–245 days, respectively for completion (Potter et al. 1980, Beamish & Thomas

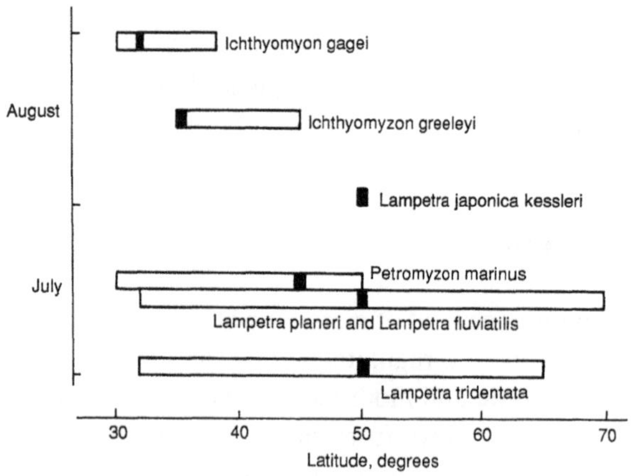

Fig. 5. Commencement of metamorphosis in relation to latitude. Open and solid bars indicate distribution range and actual location of the species studied (Poltorykhina 1971, Bird & Potter 1979a, Youson & Potter 1979, Potter et al. 1980, Richards 1980, Beamish & Thomas 1984).

1984, Bird & Potter 1979a). Curiously, there is no obvious pattern between length of time to complete metamorphosis and latitude nor do parasitic and nonparasitic species differ in a consistent pattern in their time requirements. The much greater investment to gonadal growth by nonparasitic lampreys during metamorphosis might be expected to extend the duration of metamorphosis, although this is clearly not the case for those brook lampreys which have been investigated. Rather, the time to complete metamorphosis appears to be directly related to species specific complexities in the formation of new structures modified to some degree by the environmental milieu. The shift in sex ratio during metamorphosis in favor of males may reflect a slightly longer average larval life for females (Hardisty 1961). When the age structure of *I. greeleyi* from Bent Creek was examined on July 28th, 1984, just prior to the onset of metamorphosis, males ranged in age from 1 to 4 years while females ranged from 1 to 5 years (Medland & Beamish 1987).

The adult body form which characterizes lampreys develops during metamorphosis. Patterns of change in body form are not the same for all lampreys. Even the direction of change may vary. Trunk length of *I. greeleyi* does not change appreciably throughout postlarval development (Beamish & Austin 1985) as is the case also for *I. gagei*

Table 2. Phase of occurrence of diagnostic characteristics in external metamorphosis of parasitic and nonparasitic lampreys.

	Number of phases	Pupil and iris differentiated	Oral disc complete	Oral fimbriae present	Piston distinct	Teeth and laminae with sharp hard tips	Lateral line differentiated	Reference
Parasitic								
Petromyozon marinus	7	3	4	5	5	7	6	Youson & Potter (1979)
Lampetra fluviatilis	6	3	4	6	3	6	6?	Bird & Potter (1979a)
Lampetra tridentata	7	4	4	4		7		Richards (1980)
Geotria australis	7	4	4	5	3	6		Potter et al. (1980)
Nonparasitic								
Ichthyomyzon gagei	7	4	5	5	3	7	6	Beamish & Thomas (1984)
Ichthyomyzon greeleyi	7	4	5	4	3	7	6	
Lampetra japonica kessleri	10	6–7		6–7		10		Poltorykhina (1971)
Lampetra planeri	6	3	4	6	3	6	6	Bird & Potter (1979a)

(Beamish 1982) but it declines over the same period in *G. australis* (Potter et al. 1980), *L. planeri, Lampetra fluviatilis* (Bird & Potter 1979b) and *L. tridentata* (Richards 1980). Tail length in *I. greeleyi* remains essentially unchanged throughout metamorphosis in accord with the pattern found in *L. fluviatilis* and *L. tridentata* but in contrast to the increase observed in *I. gagei*. Branchial length in all species declines during metamorphosis whereas prebranchial length increases due to the elaboration of the oral disc. Oral disc diameter of *I. greeleyi* continues to increase throughout metamorphosis as has been described also for other species. A similar pattern of increase in posterior dorsal fin height persists in all species with relative values of recently metamorphosed animals at least twice those for phase 1 individuals. Maximum eye diameter is reached in *I. greeleyi* only after the completion of metamorphosis (Beamish & Austin 1985) in accord with the pattern described for *L. planeri* (Bird & Potter 1979b).

Some changes in body form experienced by the mountain brook lamprey during metamorphosis are of value in reproduction. Others represent evidence of the ancestral origin of nonparasitic species. Enlargement of the dorsal fin and eye development are of importance to adults in maintaining their position in the fast flowing riffle areas which characterize the spawning habitat. The oral disc, which in parasitic species is used to attach to host fishes, is employed by both parasitic and nonparasitic lampreys in nest construction.

Total length and body weight of the mountain brook lamprey did not change significantly throughout metamorphosis. Surprisingly small shifts in length and weight have been recorded for other lampreys during metamorphosis (Manion & Stauffer 1970, Bird & Potter 1979b, Beamish & Thomas 1984). The general view is that lampreys do not feed during metamorphosis with the loss of body energy reserves being countered by water uptake (Moore & Potter 1976, Beamish et al. 1979). In this regard, it is interesting that Poltorykhina (1971) found the nonparasitic *L. j. kessleri* to continue to feed throughout metamorphosis. However, in the case of *I. greeleyi,* it is reasonable to assume metamorphosing animals do not feed but maintain body size by water uptake. This derives from measurements of oxygen consumption for ammocoetes of *I. greeleyi* (= *I. hubbsi*) by Hill & Potter (1970). Based on their measurements (9.5°C) and on energy concentration determined for *I. gagei* (Beamish & LeGrow 1983), the energy requirements for metamorphosis might be achieved without loss of body weight by increasing relative water content from 75 to 79–81%. Preliminary data indicate body water contents of 75.6 ± 0.5 (n = 9), 76.2 ± 0.7 (n = 14) and 76.8 ± 0.9 (n = 13) for *I. greeleyi* in phases 1, 3 and 4, respectively. By the completion of metamorphosis body water content increased to 79.4 ± 3.7% (n = 6), a value almost identical to that predicted from the earlier metabolic studies.

Acknowledgements

Gratitude for assistance is expressed to E.W. and R.B. Beamish and the North Carolina Wildlife Resources Commission. K. Melville prepared the drawings in Figures 1 and 2. Financial assistance was provided in part through an operating grant (A3218) from the Natural Sciences and Engineering Research Council of Canada.

References cited

Balon, E.K. 1980. Early ontogeny of the European landlocked arctic charr – altricial form, *Salvelinus (Salvelinus) alpinus alpinus*. pp. 607–630. *In:* E.K. Balon (ed.) Charrs: Salmonid Fishes of the Genus *Salvelinus,* Perspectives in Vertebrate Science 1, Dr W. Junk Publishers, The Hague.

Balon, E.K. 1981. Saltatory processes and altricial to precocial forms in the ontogeny of fishes. Amer. Zool. 21: 573–596.

Balon, E.K. 1983. Epigenetic mechanisms: reflections on evolutionary processes. Can. J. Fish. Aquat. Sci. 40: 2045–2058.

Balon, E.K. 1986. Saltatory ontogeny and evolution. Rivista di Biologia (Biology Forum) 79: 151–190.

Battle, H.I. & K. Hayashida. 1965. Comparative study of the intraperitoneal alimentary tract of parasitic and nonparasitic lampreys from the Great Lakes region. J. Fish. Res. Board Can. 22: 289–306.

Beamish, F.W.H. 1982. Biology of the southern brook lamprey, *Ichthyomyzon gagei*. Env. Biol. Fish. 7: 305–320.

Beamish, F.W.H. & M. LeGrow. 1983. Bioenergetics of the southern brook lamprey, *Ichthyomyzon gagei*. J. Anim. Ecol. 52: 575–590.

Beamish, F.W.H. & E.J. Thomas. 1984. Metamorphosis of the southern brook lamprey, *Ichthyomyzon gagei*. Copeia 1984: 502–515.

Beamish, F.W.H. & L.S. Austin. 1985. Growth of the mountain brook lamprey *Ichthyomyzon greeleyi* Hubbs and Trautman. Copeia 1985: 881–890.

Beamish, F.W.H., I.C. Potter & E. Thomas. 1979. Proximate composition of the adult anadromous sea lamprey, *Petromyzon marinus,* in relation to feeding, migration and reproduction. J. Anim. Ecol. 48: 1–19.

Bird, D.J. & I.C. Potter. 1979a. Metamorphosis in the paired species of lampreys, *Lampetra fluviatilis* (L.) and *Lampetra planeri* (Bloch). 1. A description of the timing and stages. J. Linn. Soc. London. Zool. 65: 127–143.

Bird, D.J. & I.C. Potter. 1979b. Metamorphosis in the paired species of lampreys, *Lampetra fluviatilis* (L.) and *Lampetra planeri* (Bloch). 2. Quantitative data for body proportions, weights, lengths and sex ratios. J. Linn. Soc. London. Zool. 65: 145–160.

Hardisty, M.W. 1961. Sex composition of lamprey populations. Nature 191: 1116–1117.

Hill, B.J. & I.C. Potter. 1970. Oxygen consumption in ammocoetes of the lamprey *Ichthyomyzon hubbsi* Raney. J. exp. Biol. 53: 47–57.

Hilliard, R.W., D.J. Bird & I.C. Potter. 1983. Metamorphic changes in the intestine of three species of lampreys. J. Morph. 176: 181–196.

Manion, P.J. & T.M. Stauffer. 1970. Metamorphosis of the landlocked sea lamprey, *Petromyzon marinus*. J. Fish. Res. Board Can. 27: 1735–1746.

Medland, T.E. & F.W.H. Beamish. 1987. Age validation for the mountain brook lamprey, *Ichthyomyzon greeleyi*. Can. J. Fish. Aquat. Sci. 44: 901–904.

Moore, J.W. & I.C. Potter. 1976. Aspects of feeding and lipid deposition and utilization in the lampreys, *Lampetra fluviatilis* (L.) and *Lampetra planeri* (Bloch). J. Anim. Ecol. 45: 699–712.

Poltorykhina, A.N. 1971. Metamorphosis of the Arctic brook lamprey [*Lampetra japonica kessleri* (Anikin)] in Upper Irtysh. J. Ichthyol. 11: 281–285.

Potter, I.C. 1980. Ecology of larval and metamorphosing lampreys. Can. J. Fish. Aquat. Sci. 37: 1641–1657.

Potter, I.C., R.W. Hilliard & D.J. Bird. 1980. Metamorphosis in the southern hemisphere lamprey, *Geotria australis* Gray. J. Zool. (London) 190: 405–430.

Richards, J.E. 1980. The freshwater life history of the anadromous Pacific lamprey, *Lampetra tridentata.* M.Sc. Thesis, University of Guelph, Guelph. 99 pp.

Ricker, W.E. 1975. Computation and interpretation of biological statistics of fish populations. Environment Canada, Fisheries and Marine Service, Bulletin 191. 382 pp.

Vladykov, V.D. & I. Kott. 1979. List of northern hemisphere lampreys (Petromyzonidae) and their distribution. Misc. Spec. Publ. 42, Dept. Fish. Oceans, Ottawa.

Youson, J.H. 1980. Morphology and physiology of lamprey metamorphosis. Can. J. Fish. Aquat. Sci. 37: 1687–1710.

Youson, J.H. & I.C. Potter. 1979. A description of the stages in the metamorphosis of the anadromous sea lamprey, *Petromyzon marinus* L. Can. J. Fish. Aquat. Sci. 57: 1808–1817.

Environmental Biology of Fishes Vol. 23, No. 1–2, pp. 55–63, 1988

Identification of lamprey larvae of the genus *Ichthyomyzon* (Petromyzontidae)

Jacqueline Lanteigne
Ichthyology Section, National Museum of Natural Sciences, Ottawa, Ontario K1A 0M8, Canada

Received 30.11.1987 Accepted 22.12.1987

Key words: Taxonomic key, Ammocoetes, Agnatha, Cyclostomata

Synopsis

This study deals with the taxonomy of ammocoetes of the North American lamprey genus *Ichthyomyzon*. The characters most useful to identify them are the pigmentation patterns on the caudal fin and the different areas of the head, as well as the number of myomeres. These characters serve as the basis for a taxonomic key to ammocoetes 56 mm in total length and up.

Introduction

The freshwater lamprey genus *Ichthyomyzon* Girard, 1858 is restricted to the tributaries of the Gulf of Mexico, the Gulf of St. Lawrence and Hudson Bay in eastern North America. The genus, whose diagnostic character is the presence of one dorsal fin, comprises six species. Until now, only the adult period has been studied in detail, mainly by Hubbs & Trautman (1937). Their revision was based on the number of trunk myomeres, the number, form and arrangement of the disc and lingual tooth series and the proportionate measurement of body parts. But neither dentition nor some of the body measurements most useful in separating the adults can be used in distinguishing the ammocoetes as they do not possess a disc. The aim of this study was to find new diagnostic features and create a taxonomic key to the ammocoetes, using modifications of the identification system developed by Vladykov (1950) for ammocoetes with two dorsal fins.

Materials and methods

The 323 specimens examined were preserved in 4–5% formalin. The abbreviation mean: AU = Auburn University, Auburn, Alabama; NMC = National Museums of Canada, Ottawa, Ontario; PSU = Pennsylvania State University, University Park, Pennsylvania; TU = Tulane University, New Orleans, Louisiana; UMMZ = University of Michigan, Museum of Zoology, Ann Arbor, Michigan; and UWSP = University of Wisconsin, Stevens Point, Wisconsin.

Ichthyomyzon bdellium

NMC 82-123; French Ck., at jct. of 6 and 6–19, 3.2 km W Mill Village, Erie Co., Pennsylvania; 22 specimens, 110–169 mm TL; May 14, 1977; J. Lanteigne, E. Kott, E.L. Cooper.

Uncatalogued; Allegheny R., at Kinzua, Warren Co., Pennsylvania; 2 specimens, 115–129 mm TL; August 9, 1960; E.L. Cooper.

PSU 131; Brokenstraw Ck., at mouth of Little Brokenstraw Ck., Warren Co., Pennsylvania; 11 specimens, 111–159 mm TL; July 27, 1965; T. Wohnsiedler and J. Nist.

Ichthyomyzon castaneus
NMC 77-1768; Rat R., 9.6 km upstream from St. Malo dam, Manitoba; 1 specimen, 152 mm TL; no collection date; V. Gyselman.
Uncatalogued; White R., between Shell Knob and Viola, Barry Co., Missouri; 1 specimen, 140 mm TL; September 2–10, 1948; E.M. Lowry.
Uncatalogued; Black R., 4.8 km N Poplar Bluff, Butler Co., Missouri; 1 specimen, 125 mm TL; July 25, 1950; C. Purkett.
Uncatalogued; Upper Black R., near Lesterville, Reynolds Co., Missouri; 6 specimens, 75–135 mm TL; October 12–15, 1948; E.M. Lowry.
Uncatalogued; Current R., 2.4 km SW Doniphan, Ripley Co., Missouri; 1 specimen, 123 mm TL; April 24–26, 1950; M.H. Patriarche.
Uncatalogued; Current R., at Powdermill Ferry, Shannon Co., Missouri; 1 specimen, 128 mm TL; July 1–4, 1949; M.H. Patriarche.
Uncatalogued; Gasconade R., 0.6 km E Hartville, Wright Co., Missouri; 1 specimen, 117 mm TL; September 10–13, 1950; E.M. Lowry.
UWSP; Prairie R., below Dudley, Lincoln Co., Wisconsin; 10 specimens, 119–165 mm TL; September 3, 1976; G. and K. Becker.

Ichthyomyzon fossor
Uncatalogued; Manistee R., Kalkaska Co., Michigan; 8 specimens, 68–96 mm TL; August 6, 1959; J.D. Hall.
Uncatalogued; Willow Ck., Simcoe Co., Ontario; 36 specimens, 79–167 mm TL; October 27–28, 1958; R. Tiede and T. Gibbons.
Uncatalogued; outlet of Willow Ck., Simcoe Co., Ontario; 2 specimens 86–162 mm TL; August 28, 1958; D. Mason.
Uncatalogued; Sauble R., Ontario; 22 specimens, 76–159 mm TL; July 3, 1958; D. Mason.
UWSP 1318; Blake Ck., Waupaca Co., Wisconsin; 5 specimens, 136–164 mm TL; June 7, 1960; G. Becker.

Ichthyomyzon gagei
AU 1082; Escambia R., NE Atmore, Escambia Co., Alabama; 2 specimens, 79–111 mm TL; March 17, 1968; J.C. Williams, F.A. Pagan, C.W. Wade.
Uncatalogued; Crossways Ck., Escambia Co., Alabama; 6 specimens, 70–84 mm TL; May 1, 1949; J.S. Dendy.
Uncatalogued; Choclafaula Ck., Macon Co., Alabama; 48 specimens, 60–147 mm TL; June 13–15, 1951; J.S. Dendy, D.C. Scott.
TU 3464; Atchafalayae R., trib. to Bayou Cocodrie, 13.4 km W Bringhurst, Rapides Co., Louisiana; 2 specimens, 111–118 mm TL; April 4, 1952, R.D. Suttkus, B. Anderson.
Uncatalogued; Big Ck., Grant Co., Louisiana; 1 specimen, 94 mm TL; January 14, 1950; R.D. Suttkus.
TU 1261; Big Ck., trib. to Little R., at Fishville, Grant Co., Louisiana; 7 specimens, 56–102 mm TL; June 14, 1950; F.R. Cagle.
Uncatalogued; Current R., 2.4 km SW Doniphan, Ripley Co., Missouri; 2 specimens, 103–120 mm TL; July 6–12, 1948; E.M. Lowry.
Uncatalogued; Current R., E of Eminence, Shannon Co., Missouri; 1 specimen, 122 mm TL; September 25–28, 1948; E.M. Lowry.
TU 3381; Little Loco Bayou, trib. to Angelina R., 7.5 km W Nacogdoches, Nacogdoches Co., Texas; 21 specimens, 100–150 mm TL; April 7, 1952; R.D. Suttkus, B. Anderson.
Uncatalogued; Unknown locality; 5 specimens, 82–148 mm TL; no date of collection or collectors.

Ichthyomyzon greeleyi
NMC 82–123; French Ck., at jct of 6 and 6–19, 3.2 km W Mill Village, Erie Co., Pennsylvania; 21 specimens, 110–160 mm TL; May 14, 1977; J. Lanteigne, E. Kott, E.L. Cooper.
Uncatalogued; Allegheny R., near Kinzua, Warren Co., Pennsylvania; 4 specimens, 113–138 mm TL; August 9, 1960; E.L. Cooper.
UMMZ 94981; French Ck., at Wattsburg, Erie Co., Pennsylvania; 1 specimen, 118 mm TL; September 19; 1931; J.R. Greeley; Paratype.

Ichthyomyzon unicuspis
Uncatalogued; Ottawa R., at Upper Duck Island, Ottawa, Ontario; 12 specimens, 90–126 mm TL; June 4, 1979; J. Lanteigne, J.M. Hanson.
Uncatalogued; Ottawa R., at Upper and Lower Duck Islands, Ottawa, Ontario; 60 specimens, 73–138 mm TL; June 7, 1979; J. Lanteigne, J.M. Hanson.

The number of myomeres is a very important taxonomic character, particularly in ammocoetes. In his key to the species of adult Quebec lampreys, Vladykov (1949) followed the method of Hubbs & Trautman (1937). More precise and consistent limits were used in the present study. The first myomere counted is the one whose anterior septum passes entirely and distinctly behind the seventh gill opening. The last myomere is the one whose posterior septum lies immediately anterior to the cloacal slit when looking at the specimen from a ventral view.

The variation in the pattern and degree of pigmentation on the upper lip, suborbital, prebranchial and suprabranchial areas and on the caudal fin was scored and defined as follows (Lanteigne, 1981):

Upper lip

++ : pigment on the lower part of the upper lip does not reach the margin of the lip, leaving a narrow white band;

+++ : pigment reaches the margin of the lower part of the upper lip.

Suborbital area

\+ : presence of a few dispersed melanophores, giving the area an irregularly spotted appearance;

++ : melanophores widely distributed, but leaving a white blotch on the cheek;

+++ : melanophores covering the whole area.

Prebranchial area

Can be either nonpigmented (−) or pigmented in the form of a prebranchial blotch (++ or +++).

Branchial area

− : presence of a nonpigmented band above the branchial openings, about 1 mm wide;

= : presence of a wide nonpigmented band above the branchial openings, about 3 mm wide.

\+ : pigmentation reaches the level of the branchial openings.

Caudal fin

\+ : pigment is restricted to the region next to the dorsal and ventral surfaces;

++ : pigment covers most of the caudal fin, except for a narrow white band along the outer margin;

+++ : pigment follows the fin rays outward to the margin of the caudal fin.

Ranges, means and 95% confidence interval of the means were calculated for the number of trunk myomeres. The pigmentation differences between the species were evaluated using Chi-square tests. Characters thus selected were used to create a taxonomic key.

Results

Pigmentation. The pigmentation patterns on the caudal fin, the prebranchial area and the different sections of the head are summarized for each species (Table 1) as percentage of occurrence and as absolute numbers. They are used as the basis for the diagnoses, the key and the description of each species. The Chi-square values and probabilities calculated from the data in Table 1 are given in Table 3. All species pairs are significantly different in at least two pigment characters.

Meristics. Table 2 is a summary of the myomere counts for each species. The mean and its 95% confidence interval are also included. All species are significantly different in mean myomere number at a probability level of 0.05, except *bdellium* and *greeleyi.*

Diagnoses and descriptions

Ichthyomyzon fossor (Fig. 1)

Ammocoetes of *fossor* can be distinguished from the other species of *Ichthyomyzon,* except *unicuspis,* by their low number of trunk myomeres (usually 50–53) and lack of pigmented lateral line organs. It can be distinguished from *unicuspis* by the following combination of characters: suborbital ar-

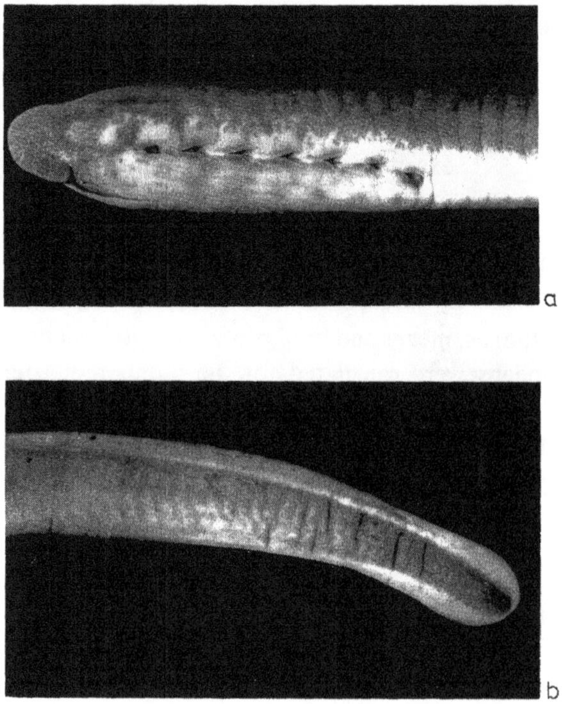

Fig. 1 (a,b). Ichthyomyzon fossor: 115 mm TL; Whitemouth R., 105 km E Winnipeg, Manitoba; May 10, 1977; H. Smart; NMC 78-0106.

ea is weakly to moderately pigmented, with melanophores evenly distributed over the area, prebranchial area is moderately to heavily pigmented with the melanophores forming a well defined spot; suprabranchial area is pigmented to within 1 mm of the branchial openings. No melanophores are present below the level of the branchial openings.

Ichthyomyzon unicuspis (Fig. 2)
Ammocoetes of *unicuspis* can be distinguished from the other species of the genus, except *fossor,* by their low number of trunk myomeres (usually 50–53) and their lack of pigmented lateral line organs. It can be distinguished from *fossor* by the following combination of characters: suborbital and prebranchial areas with pigmentation absent or weak; suprabranchial area with dispersed melanophores reaching down to the branchial openings, but a 3 mm wide band clearly visible. A few scattered melanophores are found below the level of the branchial openings.

Ichthyomyzon gagei (Fig. 3)
Ammocoetes of *gagei* can be distinguished from ammocoetes of *fossor* and *unicuspis* by the presence of pigmented lateral line organs, especially on the head and suprabranchial areas. *Gagei* larvae differ from those of *greeleyi* and *bdellium* by a lower number of trunk myomeres (usually 51–54), and from those of *castaneus,* which have a similar number of trunk myomeres, by the following combination of characters: suborbital and prebranchial areas devoid of pigment or weakly pigmented; cau-

Table 1. Variation in the degree of pigmentation on the caudal fin, the branchial area and the different sections of the head of ammocoetes of *Ichthyomyzon* (in percentage (above) and in absolute numbers (below)). Note: see Materials and methods for definition of degrees of pigmentation

Species	N	Upper lip		Suborbital area				Prebranchial area				Suprabranchial area			Caudal fin		
		++	+++	–	+	++	+++	–	+	++	+++	+	–	=	+	++	+++
I. fossor	73	76.7	23.3	–	21.9	74.0	4.1	2.7	6.9	75.3	15.1	–	91.8	8.2	30.1	67.1	2.7
		56	17	–	16	54	3	2	5	55	11	–	67	2	22	49	2
I. unicuspis	72	100	–	36.1	63.9	–	–	48.6	47.2	4.2	–	–	2.8	97.2	25.0	75.0	–
		72	–	26	46	–	–	35	34	3	–	–	2	70	18	54	–
I. gagei	95	92.6	7.4	19.0	69.5	11.6	–	41.1	55.8	3.2	–	–	100	–	45.3	54.8	–
		88	7	18	66	11	–	39	53	3	–	–	95	–	43	52	–
I. castaneus	22	50.0	50.0	–	4.5	77.3	18.2	–	4.5	68.2	27.3	45.5	54.5	–	–	95.5	4.5
		11	11	–	1	17	4	–	1	15	6	10	12	–	–	21	1
I. greeleyi	26	100	–	7.7	73.1	19.2	–	11.5	30.8	53.8	3.9	–	100	–	–	96.1	3.9
		26	–	2	19	5	–	3	8	14	1	–	26	–	–	25	1
I. bdellium	35	68.6	31.4	5.7	5.7	88.6	–	8.6	8.6	65.7	17.1	–	100	–	–	85.7	14.3
		24	11	2	2	31	–	3	3	23	6	–	35	–	–	30	5

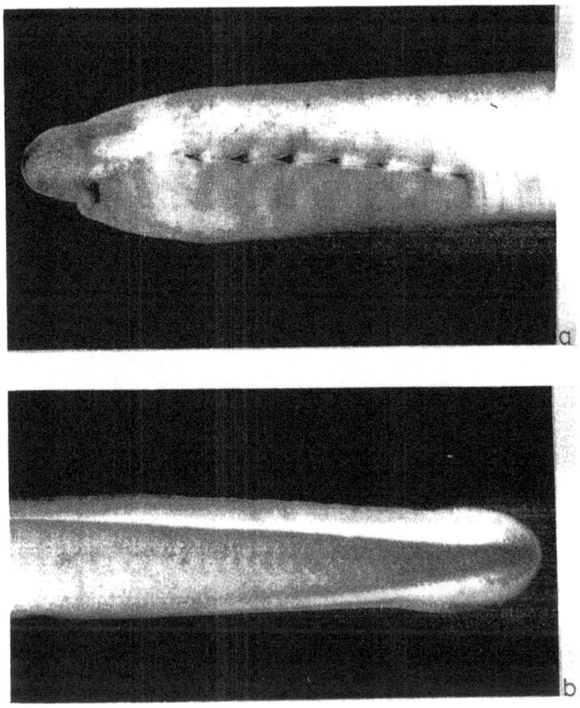

Fig. 2 (a,b). Ichthyomyzon unicuspis: 97 mm TL; Upper Duck Is., Ottawa R. at Ottawa, Ontario; June 7, 1979; J. Lanteigne, J.M. Hanson.

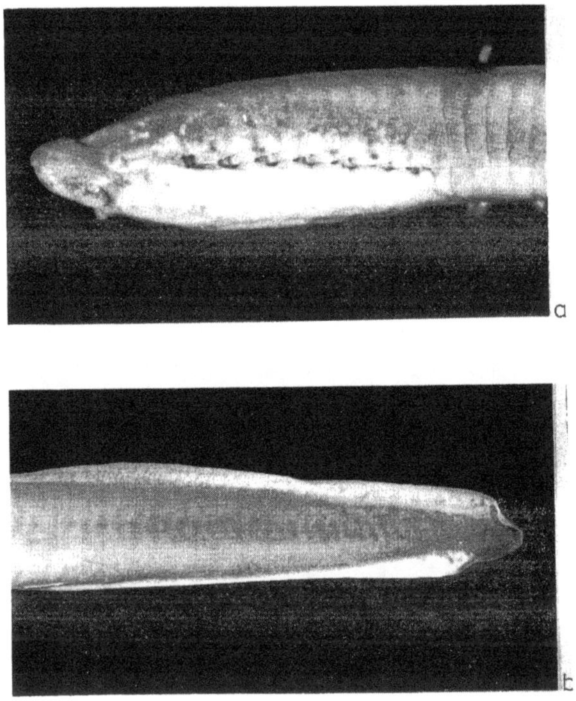

Fig. 3 (a,b). Ichthyomyzon gagei: 122 mm TL; Choclafaula Ck., Macon Co., Alabama; June 15, 1951; J.S. Dendy.

dal fin weakly to moderately pigmented.

Ichthyomyzon castaneus (Fig. 4)
Ammocoetes of *castaneus* can be distinguished from ammocoetes of *fossor* and *unicuspis* by the presence of pigmented lateral line organs, especially on the head and suprabranchial areas. It differs from ammocoetes of *greeleyi* and *bdellium* by a lower number of trunk myomeres (usually 51–54). It differs from those of *gagei*, which have a similar number of trunk myomeres, by the following combination of pigmentation characters: suborbital and prebranchial areas are moderately to heavily pigmented; caudal fin moderately pigmented.

Ichthyomyzon greeleyi (Fig. 5)
Ammocoetes of *greeleyi* can be distinguished from those of *fossor* and *unicuspis* by the presence of pigmented lateral line organs, especially on the suprabranchial area. It differs from those of *gagei* and *castaneus* in the higher number of trunk myomeres (usually 57–59). It differs from those of *bdel-*

Table 2. Myomere counts in the six species of *Ichthyomyzon* larvae (number of specimens and mean ± 95% confidence interval).

Species	N	48	49	50	51	52	53	54	55	56	57	58	59	60	Mean ± 95% C.I.
fossor	73	9	3	19	24	12	6	–	–	–	–	–	–	–	50.62 ± 0.27
unicuspis	72	1	3	8	18	23	16	3	–	–	–	–	–	–	51.65 ± 0.25
gagei	95	–	3	4	18	31	22	9	4	4	–	–	–	–	52.35 ± 0.25
castaneus	22	–	–	2	6	5	1	7	–	1	–	–	–	–	54.21 ± 0.60
greeleyi	26	–	–	–	–	–	–	–	2	1	4	14	4	1	57.77 ± 0.38
bdellium	35	–	–	–	–	–	–	–	1	9	7	12	6	–	57.37 ± 0.33

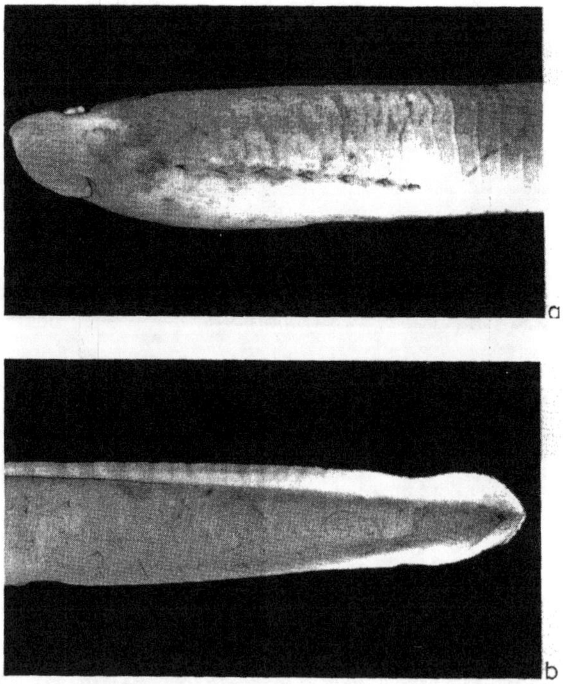

Fig. 4 (a,b). Ichthyomyzon castaneus: 144 mm TL; Prairie R., below Dudley, Lincoln Co., Wisconsin; September 3, 1976; G. and K. Becker; UWSP collection.

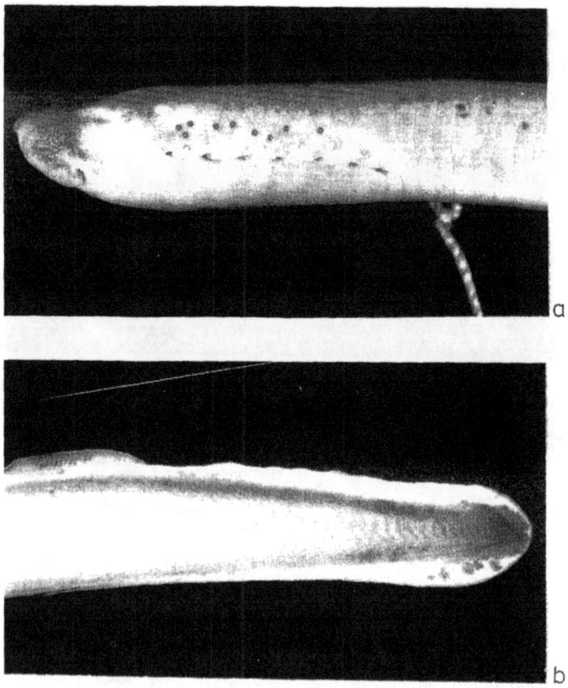

Fig. 5 (a,b). Ichthyomyzon greeleyi: 153 mm TL; French Ck., 3.2 km W Mill Village, Erie Co., Pennsylvania; May 14, 1977; J. Lanteigne, E. Kott, E.L. Cooper; NMC 82–123.

lium, which have a similar number of trunk myomeres, by the following combination of pigmentation characters: upper lip moderately pigmented; suborbital area weakly to moderately pigmented.

Ichthyomyzon bdellium (Fig. 6)
Ammocoetes of *bdellium* can be distinguished from those of *fossor* and *unicuspis* by the presence of pigmented lateral line organs, especially on the suprabranchial area. It differs from those of *gagei* and *castaneus* in the higher number of trunk myomeres (usually 56–59). It differs from those of *greeleyi,* which have a similar number of trunk myomeres, by the following combination of pigmentation characters: upper lip moderately to well pigmented; suborbital area moderately pigmented.

Discussion

The first character used for separation of the ammocoetes is the presence or absence of pigmentation of the lateral line organs. A definite dark pigmentation develops in *gagei, castaneus, bdellium* and *greeleyi.* In *fossor* and *unicuspis,* the lateral line organs are nonpigmented. This absence of pigmentation, plus the low number of trunk myomeres (usually 50–53) were therefore the criteria to distinguish the ammocoetes of the sympatric stem-satellite pair *unicuspis - fossor* from the remaining four species.

Secondary separation between the pairs with pigmented lateral line organs was achieved through the number of trunk myomeres. The sympatric stem-satellite pair *castaneus - gagei* usually has 51–54 myomeres; *castaneus* has a mean number of 54 and *gagei,* 52. The sympatric stem-satellite pair *bdellium - greeleyi* usually has 56–59; *bdellium* has 57 and *greeleyi,* 58.

Separation between members of each sympatric stem-satellite pair was achieved by using a combination of the pigmentation patterns on the caudal fin, the branchial area and the different sections of the head. The calculated Chi-square values (Table

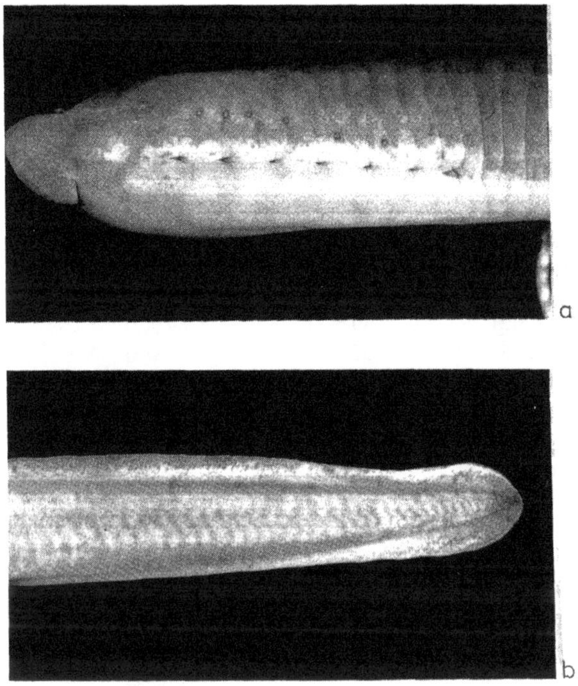

Fig. 6 (a,b). Ichthyomyzon bdellium: 162 mm TL; French Ck., 3.2 km W Mill Village, Erie Co., Pennsylvania; May 14, 1977; J. Lanteigne, E. Kott, E.L. Cooper; NMC 82–123.

3) of each pigmentation character for each pair of species demonstrated that there was a significant difference at the 95% confidence level or higher between the pigmentation patterns of each member of a pair: *unicuspis - fossor:* upper lip, suborbital, prebranchial and suprabranchial areas; *castaneus - gagei:* all pigmentation patterns; *bdellium - greeleyi:* upper lip and suborbital area. Even when ignoring the pigmentation of the lateral line organs and the number of trunk myomeres as characters to distinguish the species, it can be seen from the calculated Chi-square values of the pigmentation patterns that the ammocoetes are significantly different at the species level and can usually be distinguished on that basis. Thus, the other possible combinations of species can be separated by using a combination of the following pigmentation characters: *fossor - gagei:* all pigmentation characters; *fossor - castaneus:* suborbital and suprabranchial areas, caudal fin; *fossor - greeleyi:* upper lip, suborbital and prebranchial areas, caudal fin; *fossor - bdellium:* suborbital area, caudal fin; *unicuspis - gagei:* upper lip, suborbital and suprabranchial areas, caudal fin; *unicuspis - castaneus:* all pigmentation characters; *unicuspis - greeleyi:* suborbital, prebranchial and suprabranchial areas, caudal fin; *unicuspis - bdellium:* all pigmentation characters; *gagei - greeleyi:* prebranchial area, caudal fin; *gagei - bdellium:* upper lip, suborbital and prebranchial areas, caudal fin; *castaneus - greeleyi:* upper lip, suborbital, prebranchial and suprabranchial areas; *castaneus - bdellium:* suborbital and suprabranchial areas.

The characters most useful to separate ammocoetes at the species level are therefore the pigmentation characters on the caudal fin, the branchial area and the different sections of the head. Trunk myomeres are useful in delimiting the various stem-satellite species groups. To aid in identification of specimens, the following key was constructed.

Nonpigmented, whitish lateral line organs; myomeres 48–54 (usually 50–53)
1 2
Darkly pigmented lateral line organs, myomeres 49–60.
.... 3

Suborbital and prebranchial areas moderately pigmented; 1 mm whitish band above branchial openings.
2 .... *fossor*
Suborbital and prebranchial areas weakly pigmented; 3 mm whitish band above branchial openings.
.... *unicuspis*

Myomeres 49–56 (usually 51–54).
3 4
Myomeres 55–60 (usually 56–59).
.... 5

Prebranchial and suborbital areas weakly pigmented; caudal fin weakly to moderately pigmented.
4 .... *gagei*
Prebranchial and suborbital areas heavily pigmented; caudal fin moderately pigmented.
.... *castaneus*

Suborbital area weakly pigmented, white blotch present; upper lip moderately pigmented.

Table 3. Calculated Chi-square values, degrees of freedom (in parentheses) and probability of larval pigmentation characters of species × species of *Ichthyomyzon.*

		unicuspis	*gagei*	*castaneus*	*greeleyi*	*bdellium*
fossor	1.	18.8(1)***	8.9(1)*	5.7(1)	7.3(1)*	2.3(1)
	2.	97.5(3)***	78.4(3)***	7.4(2)*	30.5(3)***	9.8(3)*
	3.	108.6(3)***	130.1(3)***	2.3(3)	14.6(3)**	2.2(3)
	4.	115.1(1)***	8.1(1)**	84.4(2)***	2.3(1)	3.1(1)
	5.	2.6(2)	6.1(2)*	8.7(2)*	10.1(2)**	16.5(2)***
unicuspis	1.		12.5(2)**	40.8(1)***	2.8(1)	25.2(1)***
	2.		13.1(2)**	88.5(3)***	19.5(2)***	89.9(2)***
	3.		1.2(2)	74.6(3)***	37.9(3)***	69.9(3)***
	4.		159.0(1)***	37.8(2)***	88.5(1)***	98.4(1)***
	5.		7.3(1)**	6.5(1)*	10.3(2)**	19.4(2)***
gagei	1.			24.7(1)***	2.0(1)	12.3(1)**
	2.			66.8(3)***	2.5(2)	69.7(2)***
	3.			94.2(3)***	48.6(3)***	87.9(3)***
	4.			47.2(1)***	–	–
	5.			19.0(2)***	20.9(2)***	33.3(2)***
castaneus	1.				24.2(2)***	3.6(1)
	2.				28.6(3)***	7.9(3)**
	3.				11.8(3)**	2.9(3)
	4.				14.9(1)***	19.3(1)***
	5.				0.02(1)	1.4(1)
greeleyi	1.					10.0(1)**
	2.					31.9(2)***
	3.					6.9(3)
	4.					–
	5.					1.8(1)

Characters: 1– upper lip, 2– suborbital area, 3– prebranchial area, 4– suprabranchial area, 5– caudal fin
Probabilities: *: $0.01<p\leq0.05$, **: $0.001<p\leq0.01$, ***: $p\leq0.001$

5 .... *greeleyi*
Suborbital area moderately pigmented, white blotch absent; upper lip moderately to heavily pigmented.
.... *bdellium*

In their key to ammocoetes of the Great Lakes, Vladykov & Kott (1980) separated *unicuspis* from *fossor* on the basis of heavy pigmentation on the caudal fin and head in *unicuspis* and weak pigmentation in these areas in *fossor.* As shown in the present study, *fossor* has moderately pigmented suborbital and prebranchial areas, whereas *unicuspis* has weak pigmentation in those areas. The pigmentation on the caudal fin was not significantly different in the two species (Table 3). No data were presented in Vladykov & Kott's study to support their statements.

Acknowledgements

Sincere thanks to Don E. McAllister and Claude B. Renaud for their evaluation of the manuscript. The research was initially done as partial fulfillment for the M.Sc. degree at the University of Ottawa, under the guidance of Sami U. Qadri, Don E. McAllister and Vadim D. Vladykov. May Dr. Vladykov rest in peace now that this work, which constitutes the second part of his taxonomic keys of North American lamprey ammocoetes, is finally published.

References cited

Hubbs, C.L. & M.B. Trautman. 1937. A revision of the lamprey genus *Ichthyomyzon*. Misc. Publ. Mus. Zool., Univ. Michigan 35. 109 pp.

Lanteigne, J. 1981. The taxonomy and distribution of the North American lamprey genus *Ichthyomyzon*. M.Sc. Thesis, University of Ottawa, Ottawa. 150 pp.

Vladykov, V.D. 1949. Quebec lampreys. List of species and their economical importance. Dept. Fish., Quebec 26. 67 pp.

Vladykov, V.D. 1950. Larvae of Eastern American lampreys. 1- Species with two dorsal fins. Nat. Can. 77: 73–95.

Vladykov, V.D. & W.I. Follett. 1965. *Lampetra richardsoni,* a new nonparasitic species of lamprey (Petromyzonidae) from western North America. J. Fish. Res. Board Can. 22: 139–158.

Vladykov, V.D. & E. Kott. 1980. Description and key to metamorphosed specimens and ammocoetes of Petromyzonidae found in the Great Lakes region. Can. J. Fish. Aquat. Sci. 37: 1616–1625.

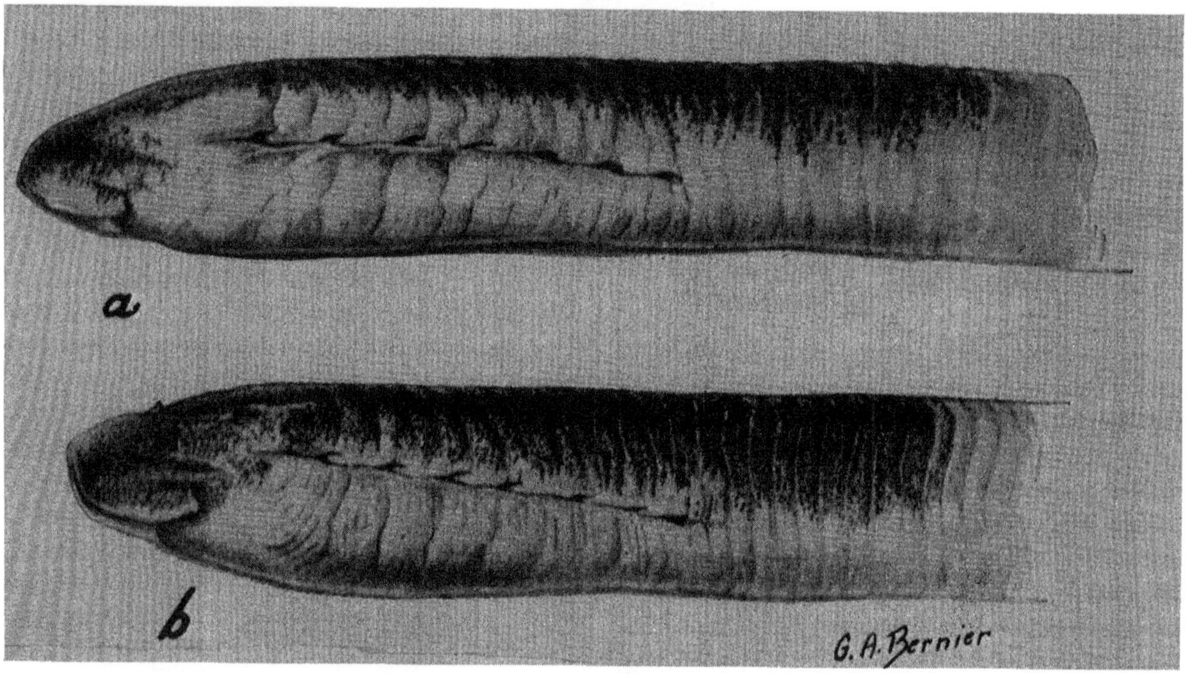

Head region of two Quebec lamprey ammocoetes: *a* – *Entosphenus lamottenii* and *b* – *Petromyzon marinus* by Germaine A. Bernier-Boulanger for Vladykov (1950).

Environmental Biology of Fishes Vol. 23, No. 1–2, pp. 65–93, 1988

Revision of *Yunnanilus* with descriptions of a miniature species flock and six new species from China (Cypriniformes: Homalopteridae)

Maurice Kottelat[1,3] & Chu Xin-Luo[2]
[1] *Laboratoire d'Ichthyologie, 2800 Delémont, Switzerland*
[2] *Kunming Institute of Zoology, Academia Sinica, Kunming, Yunnan, The People's Republic of China*
[3] *Present address: Zoologische Staatssammlung, Münchhausenstr. 21, 8000 München 60, Federal Republic of Germany*

Received 28.5.1987 Accepted 14.12.1987

Key words: Fishes, Yunnan, Burma, Endemism, Systematics, Extinction

Synopsis

The genus *Yunnanilus* Nichols, 1925 is revised; *Eonemachilus* Berg, 1938 is a junior subjective synonym. *Yunnanilus* includes at least nine described species and five undescribed species. The status of *Y. salmonides* Chaudhuri is still incertae sedis. Six new species are described: *Y. parvus, Y. altus, Y. pachycephalus, Y. niger, Y. macrogaster* and *Y. paludosus*. The last three species occur sympatrically in a small endorheic basin of eastern Yunnan; they developed different feeding specializations which allowed them to use different niches. Other species also have peculiar specializations. The diversity of feeding habits and related adaptations in *Yunnanilus* is greater than in the whole subfamily Nemacheilinae and is one more example of supralimital specialization. Speciation of fishes on the Yunnan Plateau is discussed. Several species are endangered or possibly extinct.

Introduction

Loaches of the subfamily Nemacheilinae form a very speciose group of about 450–500 nominal species, 250 of them (at least) being valid. The nomenclature of most is still very chaotic, particularly at the supraspecific level. This is partly due to their very generalized 'bauplan' with a lack of specialization which makes uneasy (or impossible) the recognition of lineages and an understanding of their phylogeny. Until now, most classifiers of nemacheilines had little choice but to rely on characters of little significance or to place most species in the catch-all genus *Nemacheilus*. Their generalized 'bauplan' has enabled nemacheilines to occupy a wide geographic area (nearly the whole Palaearctic and Oriental areas, and Tana Lake in Ethiopia) and a wide variety of habitats, including such extremes as desert springs, hot springs at 5200 m altitude in Tibet (Zhu 1981; apparently the world's highest living fishes), caves (at least 7 species; Kottelat, unpublished), or subterranean creeks (down to 400 m below surface level in Yunnan; Chu & Chen 1979).

Yunnanilus Nichols, 1925 belongs to the few easily identifiable lineages within Nemacheilinae. It can be distinguished at once by its widely separated nostrils, the anterior one being situated at the tip of a tube, and usually by the relatively (for nemacheilines) deep body.

They are known from the Yunnan Plateau, and Inlé Lake in Burma, where they usually occur in lakes, marshes or slowly flowing waters; they seem to show a particular affinity for karstic areas.

The first described species was *Nemacheilus brevis* Boulenger, 1893, from Inlé Lake, Burma. Regan (1904) described *N. nigromaculatus* and *N. pleurotaenia,* both reportedly from the lake in Yunnanfu (= Kunming). Chaudhuri (1911) described *N. salmonides* from 'Mongpan, in South Yunnan'. Nichols (1925) was the first to recognize that *N. nigromaculatus, N. pleurotaenia* and *N. salmonides* form a distinct lineage and he named it *Yunnanilus.* Berg (1938) proposed the genus *Eonemachilus* for *N. nigromaculatus,* and Jayaram (1981) referred *N. brevis* to this genus. Zhu & Wang (1985) mentioned only two species in their key to Yunnanese nemacheilines, as they considered (Zhu, personal communication) *N. salmonides* to be a junior synonym of *Y. pleurotaenia.* Zhu (1986) indicated that *Yunnanilus* includes five Chinese species, but did not provide more details. While preparing the account of nemacheilines for a handbook of Yunnan fishes, we encountered a singular species flock of 3 new species from an isolated basin in the karstic area of east Yunnan. This prompted a review of the described species and available specimens, and allowed us to discover that in fact *Yunnanilus* includes at least 14 species showing a great diversification in feeding habits and related structures. These specializations render them particularly sensitive to habitat modifications and most if not all can be considered as being threatened or endangered; one is on the verge of extinction if not already extinct.

Material and methods

Counts and measurements follow Kottelat (1984); head length is dorsal head length from tip of snout to posterior end of supraoccipital; length of caudal peduncle is from the base of last anal ray to base of lowermost principal caudal ray. Terminology for cephalic lateral line system pores follows Illick (1956).

Examined material belongs to the following institutions and collections: British Museum (Natural History), London (BMNH); Kunming Institute of Zoology, Kunming (KIZ); Muséum National d'Histoire Naturelle, Paris (MNHN); Museo Civico di Storia Naturale, Genova (MSNG); Naturhistorisches Museum, Basel (NMB); Natur-Museum und Forschungs-Institut Senckenberg, Frankfurt (SMF); Yunnan University, Kunming (YU).

As there has been much variation in the transcription of Chinese into Western languages and there are difficulties in the transcription of all sounds, we give in Figure 1 the Chinese names of localities relevant to the present text, with Pinyin transcription in the figure caption.

Yunnanilus Nichols

Yunnanilus Nichols, 1925:1 (type species *Nemacheilus pleurotaenia* Regan, 1904, by original designation). Gender: masculine.
Eonemachilus Berg, 1938:314 (type species *Nemacheilus nigromaculatus* Regan, 1904, by original designation). Gender: masculine. New synonym.

Diagnosis.
Nemacheilines with small to moderate size (3–10 cm SL). Deep to elongate body; belly flat or rounded. Dorsal fin with 4 simple and $8^1/_2$–$10^1/_2$ branched dorsal rays, with rounded supero-posterior margin, situated at about mid-length of body or slightly behind. Anal fin with 3 simple and $5^1/_2$ branched rays, its origin behind that of dorsal fin; pectoral fins with 11–13 rays, pelvic fins with 8 rays. Caudal fin emarginate, with 14 to 17 branched rays. Scales usually embedded and present on body but in some species they may be absent in predorsal area, on the belly or on the whole body except the caudal peduncle. Lateral line absent or incomplete, with a maximum of 25 pores, the posterior part of lateral line canal often split in small sections consisting of two pores and the linking canal. Lips moderately thick with almost no furrow on upper one; lower one with a median interruption and usually two deep furrows on each side. Nostrils widely separated and connected by a canal, the posterior one closer to eye than to anterior one; anterior one at tip of a tube. Cephalic lateral line pores present or absent; no pore between base of outer and inner rostral barbels; no median supra-

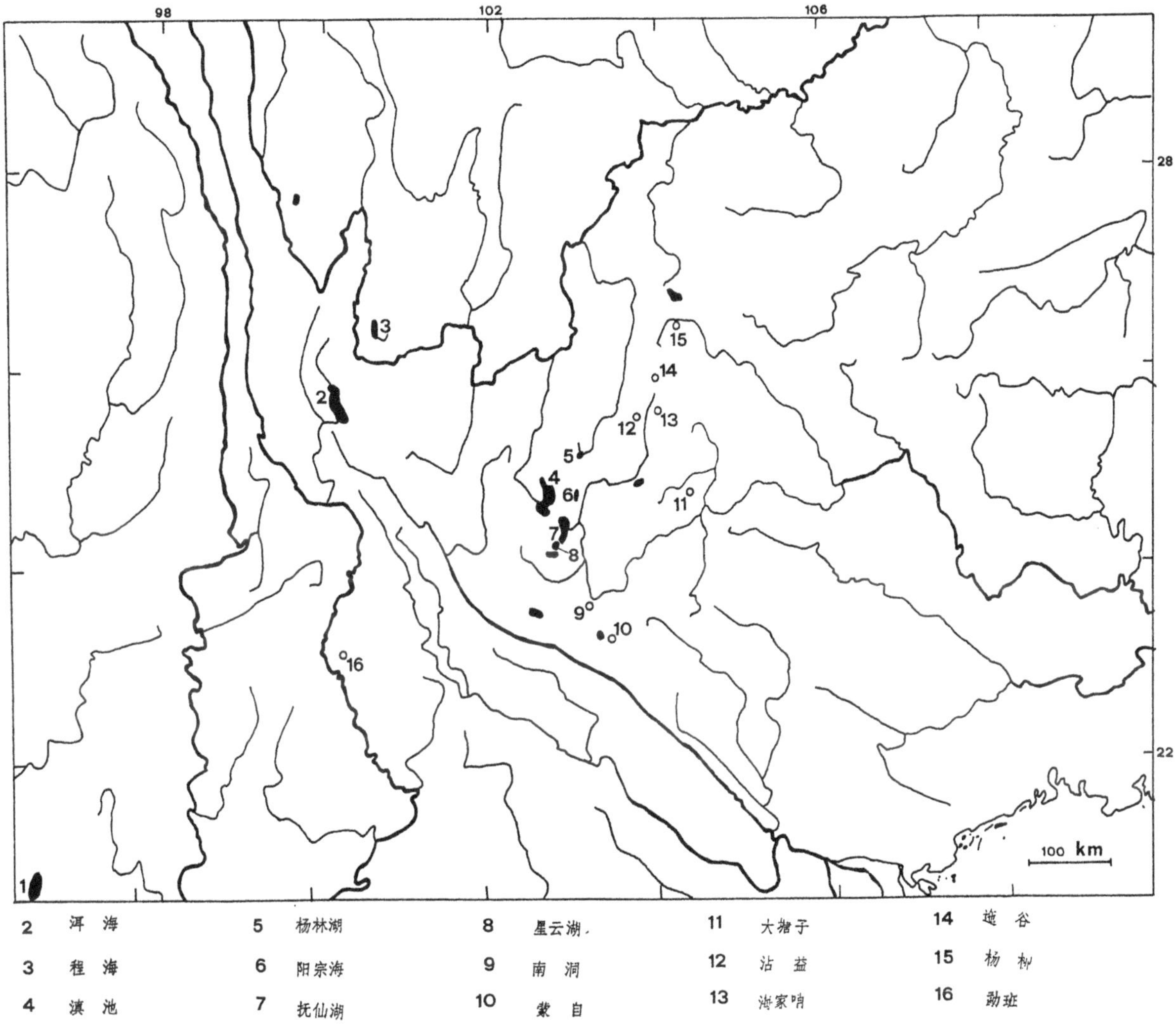

Fig. 1. Distribution of *Yunnanilus* in Yunnan and Burma and locality names in Chinese: *Y. nigromaculatus* (1), *Y.* cf. *nigromaculatus* (2, 5), *Y. altus* (13), *Y. niger* (11), *Y. pachycephalus* (15), *Y. parvus* (9), *Y. paludosus* (11), *Y. brevis* (1), *Y. macrogaster* (11), *Y. pleurotaenia* (4, 7), *Y.* cf. *pleurotaenia* (3, 5, 6, 8), *Y. salmonides* (16), *Y.* sp. Yangtzong Lake (6), *Y.* sp. Zhanyi (12), *Y.* sp. Yu Gu (14), *Y.* sp. Fuxian Lake (7), *Y.* sp. Mengzi (10), *Y.* sp. Er-Hai (2). Localities: **1** Inlé Lake, **2** Er-Hai Lake, **3** Cheng-Hai Lake, **4** Dianchi Lake, **5** Yangling Lake, **6** Yangtzong Lake, **7** Fuxian Lake, **8** Xin-Yun Lake, **9** Nan Tong Cave, **10** Mengzi, **11** Tatantze, **12** Zhanyi, **13** Haijiashao, **14** Yu Gu, **15** Yangliu, **16** Mengban.

temporal pore, the supratemporal canal is interrupted medially and has 2–4 pores on each side. Anterior chamber of swimbladder enclosed in a bony capsule; a large free posterior chamber. 2+10–13 gill-rakers. 8–14 pharyngeal teeth in a single row. Intestine short, without loop.

The distribution of the various species is shown in Figure 1.

Key to the species of *Yunnanilus*

1. – No lateral line; no cephalic sensory pores; 14 branched caudal rays. 2
   – Lateral line and cephalic sensory pores present; 15–17 branched caudal rays. 5
2. – Mouth opens forwards; mouth cleft directed

upwards; interorbital space flat or slightly concave, with a conspicuous shelf overhanging eye; dorsal profile of head with a concavity above and between eyes; Dianchi Lake. .... *Y. nigromaculatus*
– Mouth opens downwards; mouth cleft horizontal or directed downwards; interorbital space convex; dorsal profile of head without concavity above and between eyes. 3

3. – No scales on predorsal area; body black; $8^1/_2$ branched dorsal rays; Luoping Co. .... *Y. niger*
– Scales on predorsal area; body mottled, spotted or blotched; $9^1/_2$ branched dorsal rays. 4

4. – Body depth 3.5–3.9 times in SL; body mottled; Zhanyi Co. .... *Y. altus*
– Body depth 4.0–4.9 times in SL; body mottled or blotched; Xuanwei Co. .... *Y. pachycephalus*

5. – Caudal peduncle 2.0–2.3 times longer than deep; scales present only on upper and lower part of caudal peduncle; Luoping Co. .... *Y. paludosus*
– Caudal preduncle 0.9–1.5 times longer than deep; scales present on sides. 6

6. – Caudal peduncle 0.9–1.2 times longer than deep, its depth 11.3–13.4% SL; eye diameter 5.5–7.6% SL; interorbital width 7.6–10.2% SL 7
– Caudal peduncle 1.2–1.5 times longer than deep, its depth 9.3–11.3% SL; eye diameter 3.3–5.7% SL; interorbital width 6.1–7.5% SL 8

7. – Body depth 20.5–23.3% SL; 4+4 supratemporal and 11 preoperculo-mandibular cephalic lateral line pores; maximum size: 37.6 mm SL; Kaiyuan Co. .... *Y. parvus*
– Body depth 23.4–26.4% SL; 3+3 supratemporal and 9 preoperculo-mandibular cephalic lateral line pores; up to 52 mm SL; Inlé Lake in Burma. .... *Y. brevis*

Fig. 2. Yunnanilus nigromaculatus. Dianchi Lake, Yunnan, YU uncat., 72.0 and 69.5 mm SL.

8. – 6–10 pores along lateral line; preanal length 76–79% SL; head depth 13–15% SL; eye diameter 17–19% HL; body with a few eye-size spots; Luoping Co. .... *Y. macrogaster*
– 18–26 pores along lateral line; preanal length 81–83% SL; head depth 15–17% SL; eye diameter 20–29% HL; body with numerous spots or vermiculations; Dianchi and Fuxian Lakes. .... *Y. pleurotaenia*

***Yunnanilus nigromaculatus* (Regan) (Fig. 2, 3a, 4)**
Nemacheilus nigromaculatus Regan, 1904:192 original description; lake in Yunnanfu (= Kunming); syntypes: BMNH 1904.1.26: 38–39.

Material examined
BMNH 1904.1.26:38, lectotype (present designation), 58.8 mm SL; Yunnan-fu (= Kunming); Graham (paralectotype BMNH 1904.1.26:39, same data, seen after the manuscript was completed). – KIZ 01032–39, 8 ex., 49–77 mm SL; Dianchi Lake. – BMNH 1981.2.3:26–27, 2 ex., 67.0–66.8 mm SL; Kunming Shi (= Dianchi Lake); 16.7.1957. – MNHN 1906.334, 1 ex., 64.0 mm SL; Lac de Yunnan-fu; Rosenberg. – MNHN 1949.52, 3 ex., 53.2–57.8 mm SL, Kunming Lake; Cheng.

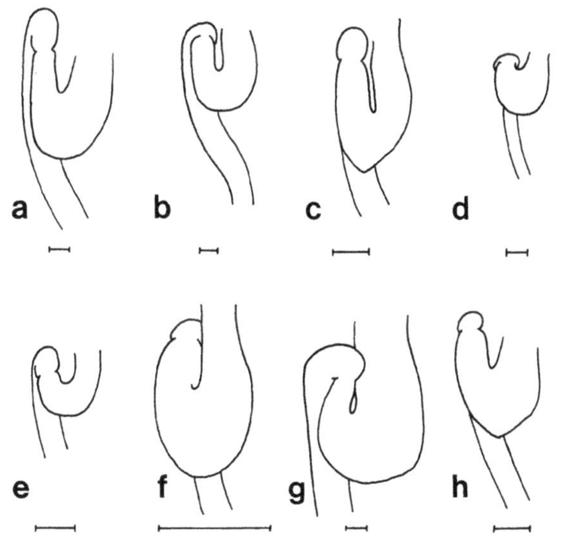

Fig. 3. Digestive tract of: a – *Yunnanilus nigromaculatus* KIZ 1086, 67.0 mm SL, b – *Y. altus* KIZ 774639, 55.5 mm SL, c – *Y. pachycephalus* KIZ 82100388, 48.2 mm SL, d – *Y. paludosus* KIZ 806061, 72.0 mm SL, e – *Y. parvus* KIZ 847177, 37.6 mm SL, f – *Y. brevis* BMNH 1930.3.3:1, 43.0 mm SL, g – *Y. macrogaster* KIZ 806064, 70.7 mm SL, h – *Y. pleurotaenia* KIZ 7811267, 57.5 mm SL. Scale bars = 2 mm.

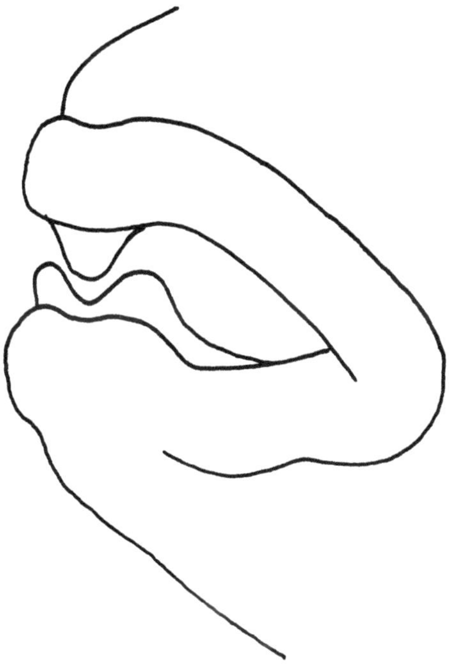

Fig. 4. Jaws of *Yunnanilus nigromaculatus,* schematized.

Diagnosis

A species of *Yunnanilus* distinguished from all other species of the genus by its terminal mouth with an upwardly directed gape. Other diagnostic characters are: absence of lateral line and head pores; scaled predorsal area; $8^1/_2$ branched dorsal rays; 14 branched caudal rays.

Description

Morphometric and meristic data are given in Table 1. A *Yunnanilus* with deep and compressed body. Dorsal profile with a strong nuchal concavity. Pectoral fin reaches about halfway to pelvic fin bases or slightly beyond. No axillary pelvic lobe but a small fold present at pelvic fin base. Pelvic fin reaches about halfway to anus which lies $^1/_2$–1 eye diameter in front of anal fin origin. Anal fin reaches caudal fin base. Caudal peduncle short, its depth 0.68–1.06 times in its length. Margin of caudal fin straight. Distal margin of dorsal fin convex.

Head compressed with flat or slightly concave interorbital area; dorsal midline with a concavity at level of eye. Nostrils widely separated; anterior one at tip of a short tube which reaches about halfway to posterior one when folded back. No pores on head. Mouth terminal, lower margin of eye level with mouth gape. Lips smooth (but all specimens examined have been preserved for a long time), upper one with a slight median incision. Upper jaw with a strong processus dentiformis, lower with a deep median notch; margin of lower jaw on side of the notch curved upwards (Fig. 4). Barbels small, all approximately equal to eye diameter. Outer rostral barbel does not reach base of maxillary barbel. Maxillary barbel reaches about front or middle of eye. No lateral line. No cephalic lateral line pores. Body entirely covered by embedded scales.

2 + 11–12 small, widely spaced and flexible gillrakers. 8–10 pharyngeal teeth. Stomach only slightly wider than intestine (Fig. 3a).

Colour pattern

Upper two thirds of body and head covered by large brown spots more or less interconnected and sometimes forming irregular vertical bars. Fins hyaline. A small vertical mark in middle of caudal fin base (not very distinct in our material).

Discussion
Zhu & Wang (1985) reported this species from Dianchi Lake and Yangling Lake (both in Jinsha-jiang basin), Yangzong Lake and Nanpang-jiang. Material from Yangtzong Lake is a distinct species that we cannot presently name and is described below as *Yunnanilus* sp. Yangzong. We have not seen material from Nanpang-jiang itself or from Yangling Lake.

BMNH 1907.5.4:74–78 (Fig. 5) is a series of three specimens, 62.5–87.3 mm SL, reported to have been collected in Tengchuan-fu (now Deng-chuan, Eryuan Co.) at the northern extremity of Er-Hai Lake and identified as *Y. nigromaculatus*. These specimens are very similar to topotypical material examined, but some differences might indicate that they represent a distinct species. They grow slightly larger (up to at least 87.3 mm SL, versus 75.9), have a deeper caudal peduncle (13.0–14.7% SL, versus 10.0–12.9; 42-51% lateral HL, versus 32–43; 1.16–1.43 times its length, versus 0.68–1.06), a deeper body (25.3–29.3 mm SL, versus 20.9–27.3). Additional material from Er-Hai is needed in order to clear the systematic status of this population. See *Yunnanilus* sp. Er-Hai, for data about Er-Hai Lake.

Habitat and biology
Although reported from other localities in Yunnan (Zhu & Wang 1985), *Y. nigromaculatus* is known definitively only from Dianchi Lake. Dianchi Lake is situated south of Kunming (25° N 102°30′ E) at an altitude of 1885 m. It is approximately 40 km long

Table 1. Morphometric and meristic data of *Yunnanilus nigromaculatus, Y. niger, Y. altus* and *Y. pachycephalus.*

	Y. nigromaculatus	*Y. niger*	*Y. altus*	*Y. pachycephalus*
N	7	1	5	7
standard length (mm)	46.9– 75.9	62.5	54.0– 70.8	46.2– 62.7
total length	117.9–126.2	119.5	116.9–121.9	117.1–122.3
dorsal head length	15.2– 19.2	18.9	18.3– 21.0	15.7– 20.8
lateral head length	27.1– 33.7	28.6	27.4– 29.0	26.7– 29.2
predorsal length	55.1– 60.8	58.1	55.1– 58.2	56.0– 60.2
prepelvic length	58.6– 65.1	58.4	56.4– 58.5	56.8– 60.0
pre-anus length	80.2– 83.6	76.0	75.0– 79.4	74.1– 80.5
preanal length	82.1– 89.1	80.6	79.1– 81.7	77.2– 83.1
length of caudal peduncle	8.9– 11.2	12.8	11.4– 13.7	10.6– 14.4
depth of caudal peduncle	10.0– 12.9	12.3	10.6– 11.8	10.7– 12.8
body depth	20.9– 27.3	26.2	25.6– 28.2	20.6– 24.9
head depth (at nape)	16.8– 22.0	19.4	17.8– 19.1	15.1– 23.3
snout length	8.2– 10.7	9.4	9.2– 10.2	8.4– 9.6
eye diameter	5.1– 6.2	4.6	4.7– 5.3	4.6– 5.6
interorbital width	8.7– 10.3	8.6	8.5– 9.5	8.2– 9.7
head width	13.2– 19.9	16.8	14.4– 20.1	14.3– 17.6
body width	9.8– 16.5	12.5	12.0– 14.2	11.3– 13.1
length of pectoral fin	12.2– 18.2	17.1	14.0– 19.8	15.2– 19.5
length of pelvic fin	11.5– 16.5	14.1	13.0– 15.9	13.1– 17.4
length of caudal fin	18.6– 23.7	18.1	18.5– 22.0	15.3– 21.2
mouth width	5.9– 10.2	7.5	6.5– 7.8	5.7– 7.9
D	$4/8^1/_2$	$4/8^1/_2$	$4/9^1/_2$	$4/9^1/_2$
A	$3/5^1/_2$	$3/5^1/_2$	$3/5^1/_2$	$3/5^1/_2$
P	11–13	12	11–12	12
V	8	8	8	8
C (branched rays)	7+7	7+7	7+7	7+7

Fig. 5. Yunnanilus cf. *nigromaculatus*. Er-Hai Lake, Yunnan, BMNH 1907.5.4:74–78, 74.1 and 62.5 mm SL.

and 12 km wide, with an area of 321 km^2 and 200 km of shore line. It is a very shallow lake, its average depth being 4.7 m and its maximum depth 6 m. Its northeastern part consists of a wide delta area. It formed in the early Pleistocene as the result of fault depressions and erosion (Yang 1984). The average annual air temperature in Kunming is 15.5° C, the minimum monthly average being 9.2° C in January and the maximum 20.1° C in July. The only available data about water temperature are for September: 18–21° C when the average air temperature is 18° C. The lake water is turbid, the transparency being 40–60 cm, the water colour 17–20 on Forel Scale, pH 8.3–8.5 and total hardness 54.5–67.2 ppm $CaCO_3$ (based in part on Ley et al. 1962). Vegetation is abundant and consists mainly of *Hydrilla verticillata* Royle, *Potamogeton malaianus* Miq., *P. crispus* L., *Najas minor* All., *Myriophyllum spicatum* L., *Ottelia alismoides* Royle and *Ceratophyllum demersum* L. (Li 1980). See Sun et al. (1986) for late Pleistocene and Holocene vegetation history, habitat modifications and photographs of the lake and surrounding areas.

Yunnanilus nigromaculatus occurs in shallow waters (less than 1 m deep) among vegetation, but not in inlets or outlets of the lake. It is a midwater fish and feeds on shrimps (all the examined stomachs contained whole shrimps only). There was originally a single species of shrimp in Dianchi Lake, *Caridina gregoriana* Kremp, which lives in dense vegetation (Tchang 1948).

Status

In the late 1950's the shrimp *Macrobrachium nipponense* (de Haan) was introduced into Dianchi Lake. It proved a very successfull competitor for the native shrimp which is now very rare (or extinct). As a result of the disappearence of its usual prey, the population of *Y. nigromaculatus* decreased quickly. To this must be added the threat posed by land reclamation for agriculture, enclosure of fish ponds, embankments, channel dredging (Sun et al. 1986:441, Fig. 11), introduction of exotic species [i.e. *Pseudorasbora parva* (Temminck & Schlegel, 1842)], agricultural, urban and industrial pollution (the lake is downstream of 1.5 million inhabitants in Kunming). As a result, *Y. nigromaculatus* is very endangered, if it is not already extinct. The last specimens we are aware of were collected in 1964.

Species originally known from the lake are: Cyprinidae: *Carassius auratus* (Linnaeus, 1758), *+*Cyprinus micristius* (Regan, 1906), *+*Cyprinus carpio chilia* Wu et al., 1963, **Anabarilius alburnops* (Regan, 1904), *+*A. polylepis* (Regan, 1904), +*Schizothorax grahami* (Regan, 1904), **Xenocypris yunnanensis* Nichols, 1925, *+*Sinocyclocheilus grahami grahami* (Regan, 1904), +*Spinibarbus sinensis* (Bleeker, 1871), +*Acrossocheilus yunnanensis* Regan, 1904, +*Discogobio yunnanensis* (Regan, 1907), *+*Acanthorhodeus elongatus* Regan, 1908; Cobitidae: *Misgurnus anguillicaudatus* (Cantor, 1842); Homalopteridae: +'*Paracobitis*' *variegatus* (Sauvage & Dabry, 1874), +*Oreias dabryi* (Sauvage, 1874), *+*Yunnanilus pleurotaenia* (Regan, 1904), *+*Y. nigromaculatus* (Regan, 1904); Siluridae: *+*Silurus mento* Regan, 1904; Bagridae: *+*Leiocassis medianalis* (Regan, 1904); Amblycipitidae: *+*Liobagrus nigricauda* Regan, 1904; Channidae: *Channa argus* (Cantor, 1842); Synbranchidae: *Monopterus albus* (Zuiew, 1793); Adrianichthyidae: *Oryzias latipes* (Temminck & Schlegel, 1846). This represents a total of 23 species, of which 12 (marked by asterisks [*]) are endemic to the lake and 16 (marked by cross [+]) are believed to be extinct or endangered in the lake (10 of them being endemics).

Yunnanilus altus new species (Fig. 3b, 6)

Holotype: KIZ 774630, 70.8 mm SL; Yunnan: Zhanyi Co.: stream Xiangshuiho at Haijiashao (25°40′N 103°57′E); April 1977.
Paratypes: KIZ 77625–629, 631–643, 650–658, 660–666, 690–696, 41 ex., 44–62 mm SL; same data.

Diagnosis
A species of *Yunnanilus* distinguished by the combination of the following characters: no pores on head; no lateral line; usually $9^1/_2$ branched dorsal rays; colour pattern consisting of minute dark spots, often fused to form irregularly shaped blotches; body depth 25–28% SL; body scaled in front of dorsal fin; 14 branched caudal rays.

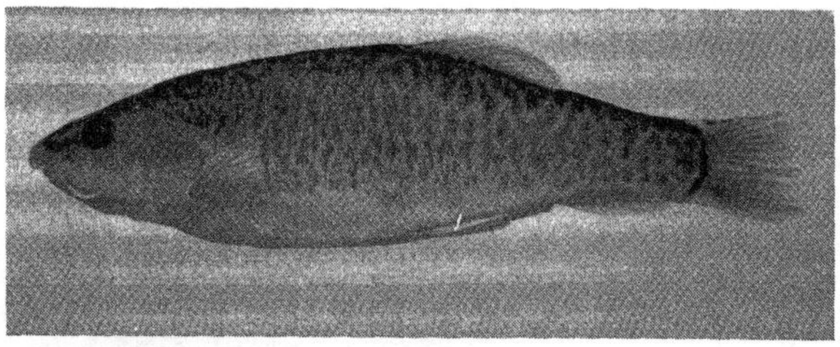

Fig. 6. Yunnanilus altus. Holotype, Zhanyi Co., Yunnan, KIZ 774630, 70.8 mm SL.

Description
Morphometric and meristic data are given in Table 1. A *Yunnanilus* with deep and compressed body. Dorsal and ventral profiles strongly convex. Pectoral fin reaches $^1/_3$–$^1/_2$ of distance to pelvic origin. No axillary pelvic lobe. Pelvic fin reaches $^1/_2$ of distance to anal fin, or slightly beyond. Anus about $^1/_2$ eye diameter in front of anal fin. Anal fin nearly reaches caudal fin base. Distal margin of dorsal fin convex. Caudal fin slightly emarginate. Caudal peduncle 1.01–1.20 times longer than deep.

Head compressed, with a convex interorbital area. Sometimes a nuchal concavity is present. Nostrils widely separated, anterior one at tip of a tube which, when folded back, reaches at most halfway to posterior one. Mouth inferior. Lower margin of eye conspicuously above level of mouth gape. Upper lip with a few shallow furrows and a median incision, lower one with a median incision and a lobe on each side. A shallow median notch on lower jaw; processus dentiformis poorly developed. Outer rostral barbel reaches beyond base of maxillary barbel, which reaches beyond vertical of hind margin of eye.

No lateral line. No cephalic lateral line pores. Body completely covered by embedded scales, except between pectoral fins.

2+12 short, conical and soft gill rakers. About 11 pharyngeal teeth. Swimbladder with a large posterior chamber. Stomach small, its diameter not conspicuously different from that of intestine (Fig. 3b), with thick walls, about $^1/_4$–$^1/_5$ of diameter; intestine with numerous finger-like projections (villi).

Fig. 7. Xiangshuiho at Haijiashao, Zhanyi Co., Yunnan; type locality of *Yunnanilus altus;* 1st January 1987.

Colour pattern
Body and head, above level of pectoral fins covered by minute dark brown spots. These spots tend to agglomerate to form irregular blotches encircled by lighter reticulations (more obvious in larger specimens). Two small dark spots at lower half of last simple dorsal ray. Black caudal basal bar continuous all along margin of hypural complex, sinuous. Two dark spots along upper margin of caudal fin, in proximal half. Other fins hyaline.

Etymology
altus (Latin): high, in allusion to the deep body.

Habitat and biology
At the type locality, the Xiangshuiho is a small stream with moderately flowing water (Fig. 7); it quickly sinks underground at 25°41′N 104°15′E. The only other species collected with *Y. altus* was *Sinocyclocheilus malacopterus* Chu & Cui, 1985. On 1st January 1987, the locality was visited again by Cui Gui-Hua and Yang Jun-Xing; the water was very clear and its temperature was 10° C. The banks are made of yellowish clay. The bottom consists of sand and pebbles, the pebbles being covered by a thick growth of algae. *Yunnanilus altus* lives in a moderate current, slowly moving in shoals on the bottom. The stomach of KIZ 774639, a ripe male 55.5 mm SL, contains sand, filamentous algae and unidentifiable vegetal remains; the intestine was empty. The stomach of KIZ 774695, a male 54.4 mm SL contains sand, unidentifiable remains and a 10 mm long insect leg which occupied the whole length of the stomach. From this it appears that *Y. altus* is detritivorous or omnivorous.

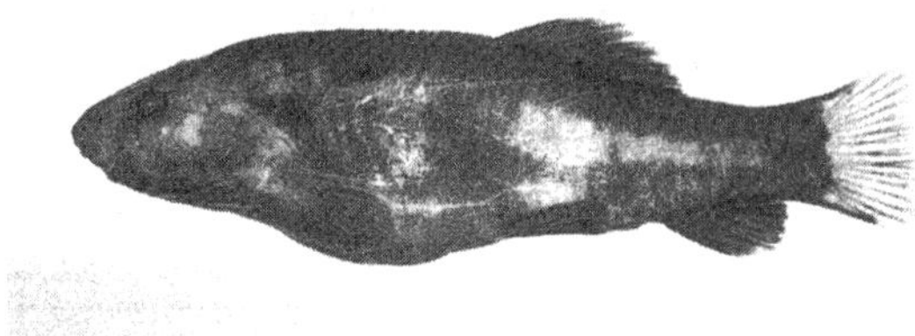

Fig. 8. Yunnanilus niger. Holotype, Luoping Co., Yunnan, KIZ 806075, 62.5 mm SL.

Yunnanilus niger **new species (Fig. 8)**

Holotype: KIZ 806075, 62.5 mm SL; Yunnan: Luoping Co.: Tatantze (24°52′N 104°18′E); Coll. Chen Yin-Rui, June 1980.

Diagnosis
A species of *Yunnanilus* distinguished by its deep black body and fins (except the caudal fin). Other diagnostic characters are: no lateral line; no head pores; no scales along dorsal mid-line in front of dorsal fin; body depth 26% SL, $8^1/_2$ branched dorsal rays; 14 branched caudal rays.

Description
Morphometric and meristic data are given in Table 1. Body deep and compressed. Dorsal profile strongly convex, ventral profile slightly convex. Pectoral fin does not reach halfway to pelvic fin base. No axillary pelvic lobe. Pelvic fin reaches slightly beyond midway to base of anal fin. Anus $^1/_2$–1 eye diameter in front of anal fin. Anal fin does not reach caudal fin base. Distal margin of dorsal fin convex. Caudal peduncle 1.04 times longer than deep. Caudal fin slightly emarginate.

Head deep, compressed, with a very slightly marked nuchal concavity. Nostrils widely separated, the anterior one at extremity of a tube which, when folded back, does not reach halfway to posterior one. No pores on head. Mouth inferior. Lips thick, without conspicuous furrows; upper one with a small median incision. Lower one with very slight median incision, the two halves nearly continuous, 1–2 transverse furrows on each side of incision forming 1–2 lobes. Processus dentiformis well developed. Lower jaw very strong. A shallow concavity in lower jaw whose anterior edges are folded upwards.

Outer rostral barbel reaches base of maxillary barbel. Maxillary barbel reaches vertical of hind margin of eye.

No lateral line. No cephalic lateral line pores. Body entirely covered by embedded scales, except between pectoral fins and on dorsal midline in front of and behind dorsal fin where the skin is thick and folded.

2+11 gill-rakers. The single specimen was not dissected, so that we cannot provide data about the digestive tract, swimbladder and stomach content.

Colour pattern

Body and head brownish black. Dorsal, anal and pelvic fins black. Caudal and pectoral fins greyish. A continuous black basal caudal bar. In life: deep velvet black.

Habitat

The holotype and only known specimen was collected in Tatantze, in Luoping Co. This area is characterized by extensive karst formation. The type locality is a small marsh, about 0.1 km^2, occupying part of a bed of a former stream. The marsh has no inlet, the water originating from rain and underwater springs of unknown origin; the outlet quickly sinks in the ground. Vegetation abundant (Fig. 9). Human pressure is important, several small villages being situated in close proximity. No data are available on water chemistry. On 12 January 1987, the locality was visited by Cui Gui-Hua and Yang Jun-Xing; they found that the spring discharged about 0.5 m^3min^{-1}; the water was very clear and its temperature was 16° C. *Yunnanilus* species were collected where the current is relatively moderate.

Fig. 9. Tatantze, Luoping Co., Yunnan; type locality of *Yunnanilus niger, Y. paludosus* and *Y. macrogaster;* 12 January 1987.

Other species present in the marsh are the introduced cyprinids *Hemiculter leucisculus* (Basilewsky, 1855), *Abbottina rivularis* (Basilewsky, 1855) and *Pseudorasbora parva* (Temminck & Schlegel, 1842). Native species are: the cyprinids *Sinocyclocheilus malacopterus* Chu & Cui, 1985 and an unidentified *Discogobio* sp. (only unidentifiable juveniles were obtained); *S. malacopterus* is widely distributed in Nanpan-jiang drainage, but always occurs close to cave inlets or outlets or inside caves (Chu & Cui, 1985); the genus has developed many endemic species in Yunnan, including a blind cave form (Chen & Chu, unpublished). *Sinocyclocheilus* feeds on bottom dwelling insects.

Yunnanilus niger is sympatric with *Y. macrogaster* and *Y. paludosus* also known only from this locality. We have no data on habitat partitioning.

Etymology

niger (Latin): black.

Yunnanilus pachycephalus new species (Fig. 3c, 10)

Holotype: KIZ 82100392, 58.5 mm SL; Yunnan: Xuanwei Co.: stream Weizhangho at Yangliu (26°38′N 104°15′E); January 1982.

Paratypes: KIZ 82100328–340, 346–350, 352–391, 393–410, 76 ex., 38–83 mm SL; same data.

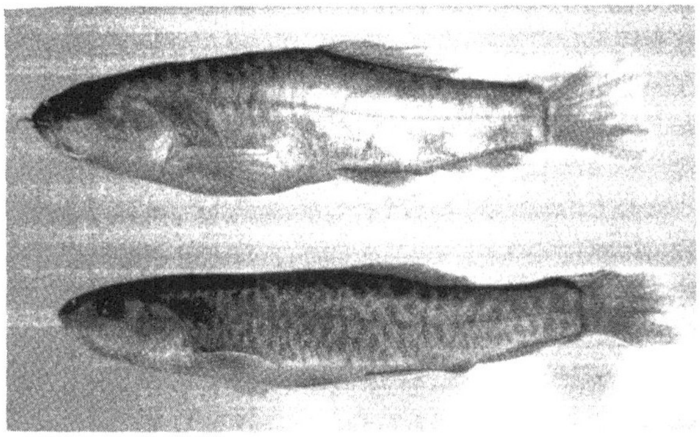

Fig. 10. Yunnanilus pachycephalus. Holotype, Xuanwei Co., Yunnan, KIZ 82100392, 58.5 mm SL and paratype, same locality, KIZ 82100359, 59.6 mm SL.

Diagnosis

A *Yunnanilus* characterized by the combination of the following characters: no lateral line; no head pores; $9^1/_2$ branched dorsal rays; 14 branched caudal rays; body depth 20–25% SL; dorsal head length 16–21% SL; colour pattern usually consisting of large dark spots.

Description

Morphometric and meristic data are given in Table 1. A moderately deep-bodied species of *Yunnanilus* with dorsal profile convex and ventral profile nearly straight. Pectoral fin reaches halfway to pelvic fin base or slightly beyond; pelvic fin reaches halfway to base of anal fin or slightly beyond (in one specimen nearly to anus). No axillary pelvic lobe. Distal margin of dorsal fin convex, posterior tip may reach above base of anal fin. Anal fin reaches to or nearly to caudal fin base. Caudal fin slightly emarginate. Caudal peduncle about as long as deep.

Head massive and wide, only slightly compressed. No nuchal concavity. Nostrils widely separated, anterior one at tip of a short barbel which does not reach halfway to posterior one when folded back. Mouth slightly inferior; lower margin of eye above level of mouth cleft. Lips thick and nearly smooth; upper one with a small median incision, lower one with a median incision and 2–3 lateral furrows not forming a conspicuous lobe. Upper jaw with a well developed processus dentiformis. Lower jaw without notch or with a very shallow concavity. Outer rostral barbel reaches base of maxillary barbel. Maxillary barbel reaches beyond vertical through posterior margin of eye.

No lateral line. No cephalic lateral line pores. Body completely covered by deeply embedded scales which are very sparsely distributed (or absent in one case) on the back in front of the dorsal fin and on the belly in front of the pelvic fins.

11 gill-rakers. 11 pharyngeal teeth. Posterior chamber of swimbladder 8 × 4 mm, its duct at least 4 mm long (in KIZ 82100337, 48 mm SL). The swimbladder is situated posterior to the stomach which is small, and whose diameter is about equal to intestine diameter (Fig. 3c).

Colour pattern

Head and body finely spotted, the spots tending to form irregular bars or larger spots, particularly along dorsal midline (this last type of colour pattern more common in specimens less than 45 mm SL). Fins hyaline. Black basal caudal bar, sometimes complete, very often much faded.

Etymology

pachy (Greek): thick, wide; *cephalus* (Greek): head; alluding to the large swollen head of some specimens.

Habitat

Stream tributary of Beipang-jiang (Beipang-jiang and Nanpang-jiang are the two main streams of the Upper Pearl River drainage). On 6 January 1987, Cui Gui-Hua and Yang Jun-Xing visited the type locality again; they report that there are high cliffs of shale on both banks and that the stream has a rocky bottom, with sandy beaches in places with a moderate current (Fig. 11). The current was swift, the water muddy and its temperature between 6 and 8° C. *Yunnanilus pachycephalus* lives in torrential waters, under rocks.

Remarks

This species shows great variation in body shape and colour pattern. We first had the impression that two species were included in the sample: one with a deep body, wide head, and a colour pattern

Fig. 11. Weizhangho at Yangliu, Xuanwei Co., Yunnan; type locality of *Yunnanilus pachycephalus;* 6 January 1987.

consisting of larger spots (Fig. 10), the other with a slender body, slender head, and a colour pattern consisting of smaller spots, thin irregular bars or uniform coloration. However, the presence of all possible intermediates seems to demonstrate the conspecifity of all our specimens.

***Yunnanilus paludosus*, new species (Fig. 3d, 12)**

Holotype: KIZ 806059, 79.9 mm SL: Yunnan: Luoping Co.: Tatantze (24°52′N 104°18′E); coll. Chen Yin-Rui, June 1980.
Paratypes: KIZ 806055–58, 60–61, 6 ex., 57.3–76.1 mm SL; same data.

Diagnosis
A *Yunnanilus* distinguished from other species by its slender and elongate body (body depth 14–16% SL) and slender caudal peduncle (1.98–2.35 times longer than deep). Other diagnostic characters are: body nearly naked, scales present only on dorsal and ventral margin of caudal peduncle; lateral line pores 16–25; head pores present.

Description
Morphometric and meristic data are given in Table 2. Body elongate and slightly compressed, its dorsal profile slowly rising up to dorsal fin base. Pectoral fin reaches about one third of distance to pelvic fin origin. No axillary pelvic lobe but a small fold is

Fig. 12. Yunnanilus paludosus. Holotype, Luoping Co., Yunnan, KIZ 806059, 79.9 mm SL.

present at pelvic fin base. Pelvic fin reaches about $^1/_2$ of the distance to anus which lies 1–1$^1/_2$ eye diameter in front of anal fin. Distal margin of dorsal fin slightly convex. Caudal peduncle slender, 1.98–2.35 times longer than deep. Caudal fin slightly emarginate.

Head with round cross section. Nostrils widely separated, space equal to eye diameter, the anterior one forming a short tube.

Lateral line short, reaching about halfway between head and dorsal origin. Cephalic lateral line system consists of 4+12 infraorbital pores, 7 supraorbital, 4+4 supratemporal, and 12 preoperculomandibular; they are difficult to count with precision as they are obliterated in most specimens. Scales sparsely distributed and embedded, on caudal peduncle, absent below and in front of dorsal fin. The largest specimens do not show any trace of squamation.

Mouth inferior. Lips not well fixed in our specimens, all of which have distended, open mouth. Lips thin, upper one apparently without furrows, and without median incision; lower with a median notch and 2–3 small lateral furrows, not forming lobes. Upper jaw with processus dentiformis. Lower jaw without notch, very thin and weak. Outer rostral barbel reaches far beyond base of maxillary one, reaching about vertical through front margin of eye. Maxillary barbel reaches to about vertical through hind margin of eye.

2+10 gill-rakers. 7 pharyngeal teeth. Stomach globular, small but its diameter distinctly larger than that of intestine (Fig. 3d); posterior chamber of swimbladder is large. The few available specimens were not dissected further.

Colour pattern
Body with 3 irregular longitudinal rows of spots. The anterior ones are vertically elongated and may be fused with those of adjacent rows. A sinuous black bar at base of caudal fin.

Habitat
See above, *Y. niger*.

Remarks
The only dissected stomach (KIZ 806058, 67.4 mm SL) contained minute flies, identified as Muscidae (Diptera) of the genus *Orthellia* by Liang Xing-Cai. This and the upwardly curved shape of the body might indicate a surface feeder (if it is not a result of fixation). The same specimen had ova about 0.8–1.0 mm in diameter. The three dissected specimens were females, and all remaining ones have swollen bellies typical of ripe females. Additional collections are needed in order to determine if males exist or if this might be an all-female species.

Etymology
palus, paludis (Latin): marsh; in allusion to the biotope where the types were collected.

Yunnanilus parvus **new species (Fig. 3e, 13)**

Holotype: KIZ 847174, 26 mm SL: Yunnan: Kaiyuan Co.: Nan Tong (23°39′N 103°17′E); coll. Cui Gui-Hua, July 1984.

Table 2. Morphometric and meristic data of *Yunnanilus pleurotaenia, Y. parvus, Y. macrogaster, Y. paludosus* and *Y. brevis*.

	Y. pleurotaenia	*Y. parvus*	*Y. macrogaster*	*Y. paludosus*	*Y. brevis*
N	19	11	3	7	4
standard length (mm)	41.9– 58.5	28.0– 37.6	66.7– 70.7	57.3– 79.9	39.7– 51.4
total length	117.8–121.7	120.3–127.1	113.3–117.3	113.5–116.2	120.8
dorsal head length	19.5– 21.9	20.0– 23.6	17.8– 19.9	15.7– 16.7	21.2– 22.9
lateral head length	23.9– 27.2	23.5– 25.8	22.6– 25.0	18.9– 23.9	25.9– 29.7
predorsal length	52.3– 59.4	51.7– 55.7	57.7– 60.1	51.9– 53.8	50.6– 55.6
prepelvic length	57.0– 60.1	55.1– 59.3	56.2– 59.1	54.1– 55.7	55.4– 59.2
pre-anus length	76.1– 78.4	72.1– 80.8	73.1– 74.4	72.5– 74.8	73.3– 77.3
preanal length	81.0– 83.4	77.0– 80.3	76.3– 78.9	75.3– 78.3	75.8– 80.4
length of caudal peduncle	12.0– 15.4	12.0– 14.6	13.6– 14.6	15.6– 18.8	12.3– 13.4
depth of caudal peduncle	9.7– 11.3	11.3– 13.1	9.3– 10.2	7.1– 8.1	11.2– 11.8
body depth	20.0– 23.2	20.5– 23.3	20.6– 22.6	14.4– 15.8	23.4– 26.4
head depth (at nape)	15.4– 16.6	15.8– 17.1	13.8– 14.4	10.5– 11.5	16.7– 18.4
snout length	8.8– 10.2	8.9– 10.1	7.6– 8.9	7.6– 8.1	8.8– 10.2
eye diameter	4.1– 5.7	5.5– 6.7	3.3– 3.5	2.9– 3.5	6.2– 7.6
interorbital width	6.1– 7.7	7.6– 10.2	7.1– 7.5	4.6– 5.8	8.4– 9.9
head width	13.1– 14.1	12.9– 15.7	13.4– 14.6	8.8– 10.4	13.8– 16.1
body width	11.2– 13.0	12.1– 15.4	12.3– 13.5	9.4– 11.1	12.3– 14.0
length of pectoral fin	14.9– 18.8	17.0– 23.6	13.5– 16.4	12.9– 13.8	17.6– 19.0
length of pelvic fin	12.6– 14.7	13.9– 17.8	9.9– 12.2	9.1– 11.0	14.4– 15.8
length of caudal fin	19.0– 22.0	23.2– 26.9	14.9– 16.5	12.1– 16.6	19.9– 20.9
mouth width	5.1– 6.2	5.1– 6.1	5.2– 6.0	3.1– 3.8	3.5– 5.8
lateral line pores	18–26	7–16	6–10	16–23	7–9
D	$4/8^{1}/_{2}$	$4/8^{1}/_{2}$	$4/8^{1}/_{2}$	$4/8^{1}/_{2}$	$4/8^{1}/_{2}$
A	$3/5^{1}/_{2}$	$3/5^{1}/_{2}$	$3/5^{1}/_{2}$	$3/5^{1}/_{2}$	$3/5^{1}/_{2}$
P	11–13	11–12	12–13	(10)11(12)	11–12
V	8	8	8	(7)8	8
C (branched rays)	8+8	8+8	8+8	8+8 or 7+8 or 7+7	8–9+8

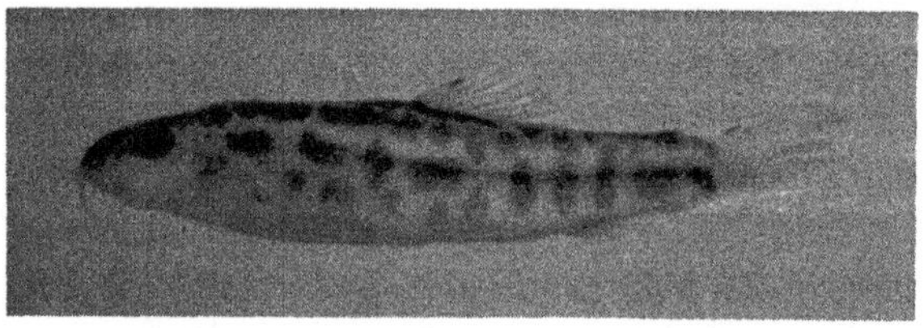

Fig. 13. Yunnanilus parvus. Paratype, Kaiyuan Co., Yunnan, KIZ 847175, 31.5 mm SL.

Paratypes: KIZ 847165–173, 175–182, 16 ex., 25–38 mm SL; same data.

Diagnosis
A dwarf *Yunnanilus* characterized by the combination of the following characters: lateral line reaching, at most, to the tip of the pectoral fins and with only 7–16 pores; cephalic lateral line pores present; caudal peduncle 0.91–1.23 times longer than deep, its depth 11–13% SL; maximum size 37.6 mm SL.

Description
Morphometric and meristic data are given in Table 2. A small, moderately deep-bodied species with blunt snout. Body compressed. Pectoral fin reaches $^1/_2$ the distance to pelvic fins or slightly beyond. No axillary pelvic lobe. Pelvic fin reaches beyond $^1/_2$ of distance to base of anal fin. Anus immediately in front of anal fin. Distal margin of dorsal fin straight or convex. Anal fin reaches base of caudal fin. Caudal fin slightly emarginate. Length of caudal peduncle 0.91–1.23 times its depth.

Head slightly compressed. Nostrils separated, distance between them about equal to half eye diameter, the anterior one tubular; when folded back the tube reaches beyond midway to posterior nostril.

Lateral line short, with 7–16 pores, reaching at most to tip of pectoral fins. Cephalic lateral line system consists usually of 4+12 infraorbital, 8 supraorbital, 4+4 supratemporal and 11 preoperculo-mandibular pores. Body completely scaled, except for belly between and in front of pectoral fins.

Mouth inferior. Lips moderately thick with a few shallow furrows. Upper lip with a very small median incision; lower one with a wide median notch and 1–2 lateral furrows forming 1–2 slightly pointed lobes. Upper jaw with processus dentiformis. Lower jaw without notch. Inner rostral barbel reaches base of maxillary barbel, outer one reaches vertical through front margin of eye. Maxillary barbel reaches anterior third of postorbital area of head.

2+12 gill-rakers. 9 pharyngeal teeth in a single row. Stomach small, its diameter about equal to that of intestine (Fig. 3e). Posterior chamber of swimbladder 4.3 × 2.0 mm (KIZ 847173, 28.9 mm SL).

Colour pattern
Body with three longitudinal rows of spots, whose number, shape and position are very variable, those in upper row often fused and connected to those of a mid-dorsal row. On middle row (along lateral line), spots are sometimes fused to form horizontally elongate spots or a stripe; spots sometimes vertically elongate, particularly the posterior ones. Lower row ending at anal fin base, often vertically elongate. Head dark on top; a dark blotch on opercle, about level with the lower row of body spots.

Dorsal fin with a dark spot at $^1/_3$ the height of last simple ray. Dark marking at base of anal fin. Caudal spot at $^1/_5$–$^1/_4$ of length of uppermost rays. Basal caudal mark consisting of two spots never forming a bar: lower one on lateral line (sometimes extenting downwards to base of lowermost branched caudal ray) and upper one at base of uppermost caudal rays.

Habitat
Nan Tong is a cave situated a few km south of Kaiyuan. A swift stream comes out from the cave all year round (Fig. 14); it immediately branches into a major, swift channel and a minor, slow channel which is diverted to ricefields. In the minor channel, there is abundant vegetation. The major channel is a tributary of Nanpan-jiang. The average annual air temperature is 19.7° C. The following species were collected with *Y. parvus* when the small channel was dried out: Cyprinidae: *Discogobio yunnanensis* (Regan, 1907); Channidae: *Chan-*

Fig. 14. Nan Tong Cave, Kaiyuan Co., Yunnan; the type series of *Yunnanilus parvus* was collected in the cave outlet; July 1984.

na asiatica (Linnaeus, 1758); Synbranchidae: *Monopterus albus* (Zuiew, 1793); Siluridae: *Silurus asotus* Linnaeus, 1758. In the major channel only the cyprinid *Acrossocheilus yunnanensis* (Regan, 1904) was collected. It is not known whether fish occur inside the cave since it cannot be entered. The examined stomachs contained only unidentifiable animal remains or detritus. KIZ 847177 is a ripe female, 37.6 mm SL, whose gonads contained ripe ova, approx. 1 mm in diameter.

Etymology
parvus (Latin): small.

Yunnanilus brevis (Boulenger) (Fig. 3f, 15)

Nemachilus brevis Boulenger, 1893:203 (original description; Inlé Lake; lectotype: BMNH 1893.6.30:83, paralectotypes: BMNH 1893.6.30: 84–85). – Annandale, 1918:43, pl. 2 fig. 1 (Inlé Lake, He-Ho Plain). – Hora, 1929:319, pl. 15 fig. 10 (Inlé Lake).
Nemacheilus brevis: Rendahl, 1948:100 (references).
Noemacheilus (Eonemachilus) brevis: Jayaram, 1981:150 (listed)
Eonemachilus brevis: Kottelat, 1986b:452 (references; colour picture of living specimen by W. Foersch).

Material examined
Burma: BMNH 1893.6.30:83, 51.4 mm SL, lectotype; Fort Stedman; Oates. – BMNH 1893.6.30: 84–85, 2 ex., 37.9–42.4 mm SL, paralectotypes; same data. – SMF 19796, 1 ex., 29.4 mm SL; about 200 km from Rangoon; coll. Tsang, don. Axelrod, V 1963. – MCSNG 29805, 1 ex., 43.2 mm SL; Inlé Lake; received from ZSI. – BMNH 1930.3.3:1, 1 ex., 43.0 mm SL; Southern Shan States; received from ZSI. – NMB 5197, 1 ex., 38.1 mm SL; Burma.

Diagnosis
A species of *Yunnanilus* characterized by the combination of the following characters: short lateral line present, with up to 12 pores; cephalic lateral line system present with 3+3 supratemporal and 9 preoperculo-mandibular pores; 16–17 branched caudal fin rays; caudal peduncle 1.06–1.14 times longer than deep, its depth 12–14% SL; eye diameter 6.2–7.6% SL; body depth 23–26% SL.

Description
Morphometric and meristic data are given in Table 2. A deep-bodied *Yunnanilus* with body depth increasing to dorsal fin origin, then decreasing; dorsal profile of caudal peduncle horizontal. Head and body compressed. Pectoral fin reaches approximately $^1/_2$–$^2/_3$ of distance to pelvic fin base. A small axillary pelvic lobe. Pelvic fin nearly reaches anus which lies approximately one eye diameter in front of anal fin. Caudal fin emarginate. Dorsal margin of dorsal fin convex. No adipose crests or keels on caudal peduncle. Caudal peduncle 1.06–1.14 times longer than deep.

Anterior nostril at the tip of a tube which reaches posterior nostril when folded back. Mouth arched, with thin lips. Upper lip with a median notch variously marked and a few furrows forming quite wide 'costae'. Lower lip with a median interruption and a few deep furrows forming lobe-like extensions on the lip. Processus dentiformis present, but not very pronounced. No median notch on lower jaw. Inner rostral barbel reaches corner of mouth, outer one reaches the vertical through middle of eye. Maxillary barbel reaches middle of postorbital area of head.

Lateral line short, with about 12 pores. Cephalic

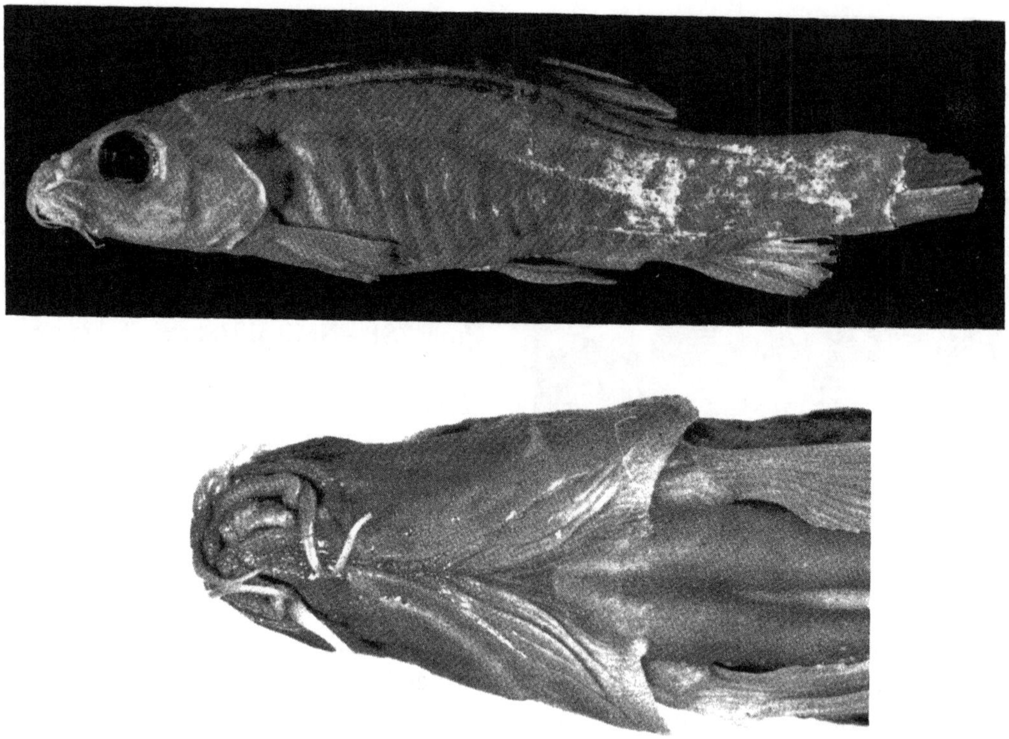

Fig. 15. Yunnanilus brevis. Inlé Lake, Burma, BMNH 1930.3.3:1, 43.0 mm SL.

lateral line system consists of 4+11 infraorbital, 7 supraorbital, 3+3 supratemporal and 9 preoperculo-mandibular pores. Body completely covered by scales which are not embedded.

Stomach diameter larger than that of intestine (Fig. 3f); in the only (and previously dissected) specimen examined for this character, the intestine had a bend somewhat behind the stomach, but the whole digestive tract was in a bad state and the bend might have been an artefact or the result of a former examination.

Sexual dimorphism

According to Annandale (1918), the males have a number of short black vertical bars on the sides, which sometimes fuse together to form an irregular stripe. The bars are variable both in number and in size. Males have a suborbital flap.

Colour pattern

The whole body is mottled or spotted. The spots are slightly vertically elongated. There is a black spot at mid-height of caudal fin base. Fins hyaline.

Distribution and habitat

Only known from Inlé Lake (Fig. 16) and the nearby He-Ho plain. Inlé Lake has been described and illustrated by Annandale (1918a, b) and Kottelat (1986a, b); it is situated in the Southern Shan States of Burma, near Taunggyi; its altitude is about 1000 m above sea level. It occupies the floor of a valley between two north-south directed mountain ranges. Like most of the Shan States, this area is rich in limestone, thus in karstic formations. According to Annandale (1918a) the outlet of Inlé Lake 'disappears into the ground, but at some considerable distance south of the lake' (p. 2). It is supposed that its waters finally reaches a tributary of the Salween. Inlé Lake is considered to be the last remnant of a lake country which existed there,

Fig. 16. Inlé Lake,Burma; type locality of *Yunnanilus brevis.*

probably until the Pleistocene; according to Bender (1983) 'they disappeared as a result of karstification and subsurface run-off in the carbonate rocks, or locally as a result of peat formation and silting-up processes'. In more or less remote times, its level was considerably higher than now.

Inlé Lake is about 20 km long and 6 km wide; its greatest depth, during the dry season (around March) is about 4 m, the average depth being about 2 m; at the end of the rainy season, the average depth is about 6 m. In fact, it is difficult to give exact dimensions as the shores are formed by a wide marshy area and floating islands. In the open area of the lake, the bottom is entirely covered by *Ceratophyllum* which grows nearly to the surface. The only data on water chemistry are those published by Annandale (1918a): total solids 0.171 g l^{-1}, organic matter 0.016 g l^{-1}, calcium 0.22 g l^{-1}, magnesium 0.279 g l^{-1}, chlorine 0.0017 g l^{-1}, sulphate 0.0017 g l^{-1}, silica 0.001 g l^{-1}, carbonic acid 0.103 g l^{-1}. The water is very clear and everywhere it is possible to see the bottom. Water temperature in March is 22° C, while the average annual air temperature is 23° C.

The following 30 other species are known from Inlé Lake and its basin (the 14 endemics are marked by an asterisk *): Cobitidae: *Lepidocephalichthys berdmorei* Blyth, 1860; Homalopteridae: *Acanthocobitis botia* (Hamilton, 1822), **Nemacheilus brunneanus* Annandale, 1918, **N. rivulicola* Hora, 1929, **N. shanensis* Hora, 1929; Cyprinidae: **Sawbwa resplendens* Annandale, 1918, **Microrasbora rubescens* Annandale, 1918, *'*M.*'*erythromicron* Annandale, 1918, **Inlecypris auropurpureus* (Annandale, 1918), **Cyprinus intha* Annandale, 1918, **Garra graveleyi* Annandale, 1919, *Cirrhinus burmanicus* Hora, 1936, *Puntius sarana* (Hamilton, 1822), **P. schanicus* (Boulenger, 1893), **P. compressiformis* (Cockerell, 1913), *P. ticto* (Hamilton, 1822), *Neolissochilus dukai* (Day, 1878), **Neolissochilus nigrovittatus* (Boulenger, 1893), *Tor tor* (Hamilton, 1822), *Barilius barnoides* (?) Vinciguerra, 1890, *Danio aequipinnatus* (M'Clelland, 1839); Clariidae: *Clarias batrachus* (Linnaeus, 1758); Notopteridae: *Notopterus notopterus* (Pallas, 1769); Channidae: *Channa striata* (Bloch, 1797), *Channa gachua* (Hamilton, 1822); Synbranchidae: *Monopterus albus* (Zuiew, 1793), *M. cuchia* (Hamilton, 1822) Chaudhuriidae: *Chaudhuria caudata* Annandale, 1918; Mastacembelidae: **Mastacembelus oatesi* Boulenger, 1893, **Macrognathus caudiocellatus* (Boulenger, 1893).

Yunnanilus macrogaster new species (Fig. 3g, 17)

Holotype: KIZ 806062, 70.5 mm SL; Yunnan: Luoping Co.: Tatantze (24°52′N 104°18′E); coll. Chen Yin-Rui, June 1980.
Paratypes: KIZ 806063–64, 2 ex., 66.7 & 70.7 mm SL; same data.

Diagnosis
A species of *Yunnanilus* distinguished by the combination of the following characters: lateral line short, with 6–10 pores; cephalic lateral line pores present; $8^{1}/_{2}$ branched dorsal rays; caudal peduncle 1.43–1.45 times longer than deep; eye diameter 17–19% dorsal HL; colour pattern consisting of a few large spots (about size of eye).

Description
Morphometric and meristic data are given in Table 2. A moderately elongated species with compressed body. Pectoral fin reaches about $^{1}/_{3}$ of distance to pelvic fin base. No axillary pelvic lobe. Pelvic fin reaches between $^{1}/_{2}$ of distance to anus to $^{1}/_{2}$ of distance to origin of anal fin. Anus 1–$1^{1}/_{2}$ eye

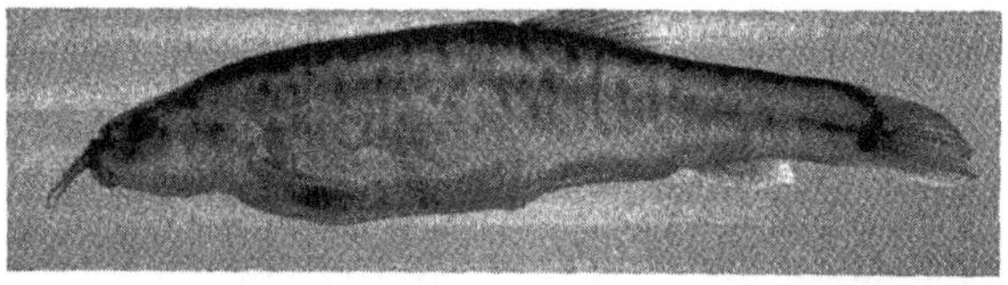

Fig. 17. Yunnanilus macrogaster. Paratype, Luoping Co., Yunnan, KIZ 806063, 66.7 mm SL.

diameter in front of anal fin origin. Distal margin of dorsal fin convex. Anal fin does not reach base of caudal fin. Caudal fin slightly emarginate. Length of caudal peduncle 1.43–1.45 times longer than its depth.

Head compressed, snout pointed when seen from the side, square when seen from above. Nostrils separated, the anterior one tubular. When folded back the tube reaches or nearly reaches posterior nostril. Mouth sub-inferior, moderately arched, upper jaw projecting beyond lower jaw. Mouth cleft extends to the vertical through the nostrils. Upper lip smooth, without median notch. Lower lip with a median interruption and a few deep furrows forming small lobes. Processus dentiformis present. No notch in the middle of lower jaw. Inner rostral barbel reaches vertical through anterior margin of eye; outer one reaches vertical through posterior margin of eye or beyond. Maxillary barbel reaches middle of postorbital area of head.

Lateral line short, restricted to 6–10 pores, not even reaching mid-length of pectoral fins. Cephalic lateral line system consists of 9–10 supraorbital, 4+13–15 infraorbital, 3+3 supra-temporal and 14–15 preoperculo-mandibular pores. Body completely scaled, including belly between pectoral fin bases.

Posterior chamber of swimbladder 14.6 × 5.7 mm (KIZ 806064, 70.7 mm SL). Stomach greatly swollen, measuring about 11 × 8 mm (same specimen) (Fig. 3g). 2+13 very short gill-rakers. 14 pharyngeal teeth.

Colour pattern

Body yellowish brown with irregularly distributed dark brown spots on its upper two-thirds; size of spots about equal to eye diameter; spots more closely set along back. Top of head darker, a dark blotch on opercle behind eye. Fins hyaline. Black basal caudal bar vertical but not reaching upper and lower margin of caudal peduncle.

Habitat

See *Y. niger*. The large stomach of KIZ 806064, 70.7 mm SL, contained four lumbricids (about 10 mm long), about 50 oligochaete worms and whole and partial larvae of Homoptera and Coleoptera (Dysticidae ?).

Etymology

macro (Greek): large; *gaster* (Greek): stomach; an allusion to the large swollen stomach.

Yunnanilus pleurotaenia (Regan) (Fig. 3h, 18, 19)

Nemacheilus pleurotaenia Regan, 1904:192 (original description; 'Sea of Tien' [= Dianchi Lake]).
? *Nemachilus salmonides* Chaudhuri, 1911:18, pl. 1 fig. 3–3a (original description; Mongpan, Southern Yunnan).

Material

BMNH 1904.1.26:36, lectotype (present designation), 43.5 mm SL; Yunnan: Yunnan-fu [= Kunming]; Graham. – KIZ 7811258–272, 15 ex., 49–60 mm SL; Fuxian Lake; November 1978. – KIZ uncat., 15 ex., 27–48 mm SL; Fuxian Lake.

Diagnosis

A species of *Yunnanilus* distinguished by the combination of the following characters: cephalic head pores present; lateral line present, with 18–26 pores; caudal peduncle 1.19–1.40 times longer than deep; $8^1/_2$ branched dorsal rays; preanal length 81–83% SL; eye diameter 20–29% HL.

Description

Morphometric and meristic data are given in Table 2. A *Yunnanilus* with moderately compressed body. Dorsal profile convex, ventral profile nearly straight. Pectoral fin reaches at most halfway to pelvic fin base. No axillary pelvic lobe. Pelvic fin

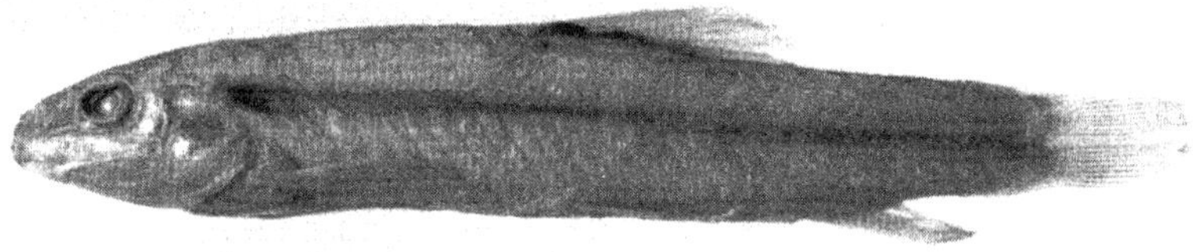

Fig. 18. Yunnanilus pleurotaenia. Lectotype, Dianchi Lake, Yunnan, BMNH 1904.1.26:36, 43.5 mm SL.

reaches about halfway to anal fin origin. Anus about one eye diameter in front of anal fin. Distal margin of dorsal fin slightly convex. Anal fin does not reach caudal fin base. Caudal fin slightly emarginate. Caudal peduncle 1.19–1.40 times longer than deep (the highest values are encountered in the largest specimens).

Head slightly compressed. Interorbital area convex. A nuchal concavity present in largest specimens. Nostrils widely separated, the anterior one at tip of a tube which, when folded back, reaches about halfway to posterior one. Mouth inferior. Lips with very few fine furrows and covered by minute papillae. A very slight median incision in upper lip. Lower lip with a wide median notch and 2 deep lateral furrows forming two lobes. Upper jaw with processus dentiformis not very developed; lower jaw entire or with a shallow median concavity but not notched. Inner rostral barbel reaches base of maxillary barbels. Outer rostral barbel reaches somewhat beyond base of maxillary one. Maxillary barbel reaches slightly beyond vertical through posterior margin of eye.

Lateral line reaches, at most, vertical of dorsal fin origin, with 12–26 pores. Cephalic lateral line pores: usually 4+12 infraorbital, 7–8 supraorbital, 3+3 supratemporal and 11 preoperculo-mandibular pores. Body and belly, including area between pectoral fin base, entirely covered by embedded scales.

Posterior chamber of swimbladder voluminous, 11 × 5.5 mm, duct 4.5 mm (KIZ 7811219, 54.0 mm SL). 2+10 gill-rakers. About 9–10 pharyngeal teeth in a single row. Stomach diameter slightly larger than that of intestine (Fig. 3h).

Colour pattern

Body entirely (except belly which is white) covered by large irregularly shaped dark brown blotches on a light brown or yellowish back ground. Fins hyaline. A black spot at caudal fin base, from mid-height to lower margin of caudal peduncle, and a smaller one at base of 3–4 uppermost caudal rays.

Discussion

Yunnanilus pleurotaenia has the widest reported distribution of any species in the genus. Zhu & Wang (1985) listed the following localities: Dianchi Lake, Yangling Lake, Cheng Hai Lake (26°27′N 100°40′E) [all in Jinsha-jiang basin], Yangzong Lake, Xin Yun Lake (24°20′N 102°47′E), Fuxian Lake [in Nanpang-jiang basin] and Er-Hai Lake [Lancang-jiang basin]. Of these we could examine material from Dianchi Lake (the lectotype) and Fuxian Lake; the above description is based on Fuxian material. Material referred to as *Y. pleurotaenia* from Er-Hai is discussed below as *Yunnanilus* sp. Er-Hai; material from Cheng Hai Lake possibly is close to it.

We could examine only a single specimen from Dianchi, the lectotype. It agrees well with our Fuxian Lake material; minor points of disagreement, probably due to preservation, and its long period of preservation, are body depth (17.9% SL versus 20.0–23.2), head depth (14.5% SL versus 15.4–16.6) and head width (11.7% SL versus 13.1–14.1). Its colour pattern is no longer distinct (Fig. 18). A collection of fresh material from Dianchi Lake or its close vicinity would be desirable in order to confirm the identification of our material. The last collection from Dianchi Lake was in 1965 and we do not know if the species is still present there.

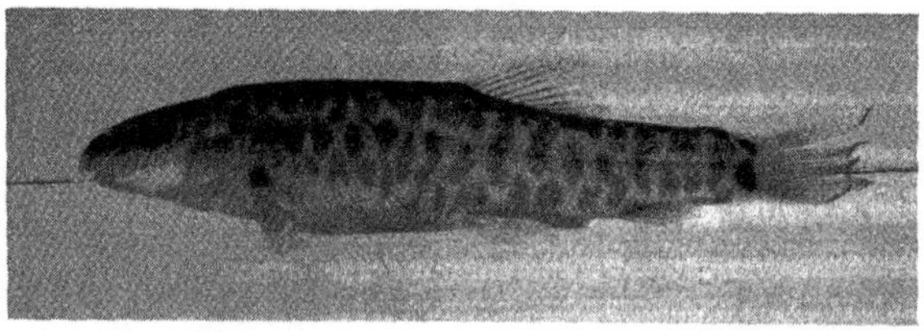

Fig. 19. Yunnanilus pleurotaenia. Fuxian Lake, Yunnan, KIZ 7811270, 42.0 mm SL.

The first author tried to borrow the holotype and only known specimen of *Nemacheilus salmonides* Chaudhuri, 1911, which is housed in Zoological Survey of India, Calcutta, but he never received an answer to his request. The exact status of this specimen can only be established by direct examination, especially since Berg (1938) has already mentioned discrepencies between Chaudhuri's original description and illustration. Until that becomes possible, we tentatively consider it to be a junior synonym of *Y. pleurotaenia.* The presence of 25 pores in the lateral line, a caudal peduncle depth about 9.5% SL (from Chaudhuri's illustration) and about 1.2 times in its length, an eye diameter 4.4% SL, and a preanal length 82% SL are diagnostic for *Y. pleurotaenia* (see key). The type locality mentioned by Chaudhuri (1911) is Mongpan. According to Coggin Brown (1910) Mong Pan is situated 23°07′N 100°23′E; following to the now official transcription system its spelling is Mengban; it is located in the Lancan-jiang (Mekong) basin. *Yunnanilus salmonides* is mentioned by Ley et al. (1963) from Er-Hai Lake, but without evidence. We believe that Ley et al. had material of the species listed below as *Y.* sp. 'Er-Hai'.

Habitat and biology

The type locality, Dianchi Lake, was described above (see under *Y. nigromaculatus*). Our material originates from Fuxian Lake. Fuxian Lake has an elongate shape, is 31.4 km long and its greatest width is 11.7 km, the average width being 7.2 km; its area is 196 km^2. Fuxian Lake formed in early Pleistocene as the result of fault depression (Yang 1984). It is very peculiar when compared with all other Yunnan lakes which are shallow (less than 20 m) by its steep profile and great depth, maximum depth being 160 m, average depth 88 m. Water temperature is between 14.9 and 24.9° C (depth not indicated), and pH between 8.1–8.7 (based on Ley et al. 1963).Data on air temperature in Ley et al. (1963) seems unreliable. We have no data about stratification, or oxygen concentration at the bottom of the lake. The deep water fauna seems completely unknown; if there is any deep water life in the lake, then other endemic *Yunnanilus* species might have evolved there.

KIZ 7811267 is a ripe female, 57.5 mm SL whose gonads contained eggs near ovulation, about 0.8 mm in diameter. Its stomach, about 6 mm long, contained a single gammarid crustacean about 6–7 mm long.

Undescribed species

Under this heading, we briefly mention various specimens mostly examined by the first author in Kunming Fisheries Research Institute and in Department of Biology, Yunnan University, Kunming. Unfortunately, time limitations, technical problems and administrative reasons did not allow detailed description and naming of these species. We hope that new collections will soon make formal description of these species possible. In addition, the first author has photographs of some other populations which might deserve species status. Regretably, too little information can be obtained from these pictures. Lastly, a species of *Yunnanilus,* of which we had only poorly preserved specimens, also occurs in Er-Hai Lake (Dali Lake) (see discussion of *Y. nigromaculatus*). Although the pictures of most of these species are not of very high quality, we considered them worth publishing in order to facilitate future identification.

Yunnanilus sp. Yangtzong Lake (Fig. 20)

The first author has seen a single specimen of this species in Kunming Fisheries Research Institute and briefly examined two specimens in the Department of Biology, Yunnan University, Kunming

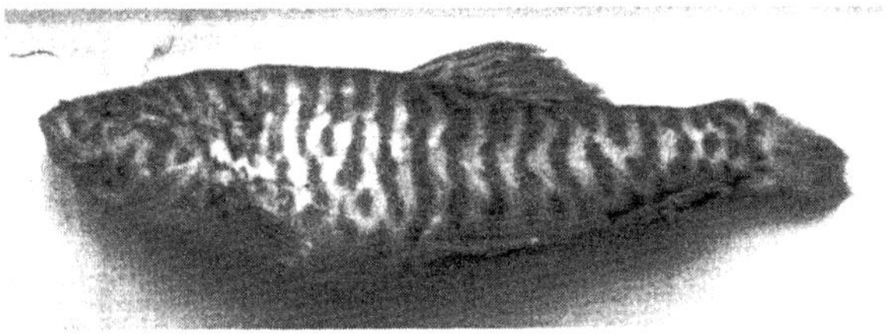

Fig. 20. Yunnanilus sp. Yangtzong Lake, Yunnan, 70.0 mm SL.

(YU 786052). Yangtzong Lake is listed as part of the distribution of *Y. nigromaculatus* by Zhu & Wang (1985). This species closely ressembles *Y. nigromaculatus* from which it is distinguished by having $9^1/_2$ branched dorsal rays (versus $8^1/_2$ in *Y. nigromaculatus*), shorter lateral head length (25.6–26.8% SL versus 27.1–33.7), shorter predorsal length (51.7–53.1% SL versus 55.1–60.8), shorter prepelvic length (54.7–57.2% SL versus 58.6–65.1), shorter pre-anus length (73.6–74.8% SL versus 80.2–83.6), shorter preanal length (79.1–79.8% SL versus 82.1–89.1), a longer caudal peduncle (13.8–14.0% SL versus 8.9–11.2, and 1.34–1.35 times longer than deep versus 0.68–1.06).

We have no doubt that this represents an unnamed species. All the available material is poorly preserved and additional collections are urgently needed as parts of Yangtzong Lake are already heavily polluted by a coal burning electricity plant. Yangtzong Lake is part of the Nanpan-jiang drainage and is situated 1770 m above sea level and measures about 12.2 × 3.0 km, with an area of 30.75 km^2. Its maximum depth is 28 m and the average depth is 20 m.

After this manuscript was completed, we were informed that this species will be described by Zhu & Cao (unpublished) as a subspecies of *Y. nigromaculatus*. We consider it worthy of species rank.

Yunnanilus sp. Zhanyi Co. (Fig. 21)

The first author briefly examined a series of specimens from Zhanyi Co. (more precise locality data are not available) in Kunming Fisheries Research Institute. They are characterized by the absence of lateral line and cephalic pores, and by a deep body

Fig. 21. Yunnanilus sp. Zhanyi Co., Yunnan, specimen on top 41.0 mm SL.

(25–30% SL). They are apparently related to *Y. altus* described above from the same county but are distinguished by having only $7^1/_2$ branched dorsal rays (versus $9^1/_2$). Females are ripe at 35 mm SL and less, while ripe *Y. altus* females are usually larger (up to 71 mm SL). Material with exact locality data would be desirable for comparison and for detailed examination.

Yunnanilus sp. Yu Gu (Fig. 22)

The first author photographed and examined a single specimen 55 mm SL from Yu Gu in Kunming Fisheries Research Institute which is apparently an undescribed species. It is distinguished by a slender caudal peduncle (15.4% SL, 1.75 times longer than deep; measurements from photograph) and a lateral line nearly reaching to the vertical through the dorsal fin origin. Among named *Yunnanilus* species, this combination of characters is not known. *Yunnanilus paludosus* has a slender caudal peduncle (2.0–2.3 times longer than deep), a slender body (14.4–15.8% SL versus 19.2) and a smaller eye (2.9–3.5% SL versus 4.5). Other species with cephalic pores and a lateral line have a stouter caudal peduncle (0.9–1.5 times longer than deep) and deeper body (20.3–23.3% SL versus 19.2); *Y. parvus, Y. macrogaster* and *Y. brevis* have shorter

Fig. 22. ***Yunnanilus*** sp. Yugu, Yunnan, 55.0 mm SL.

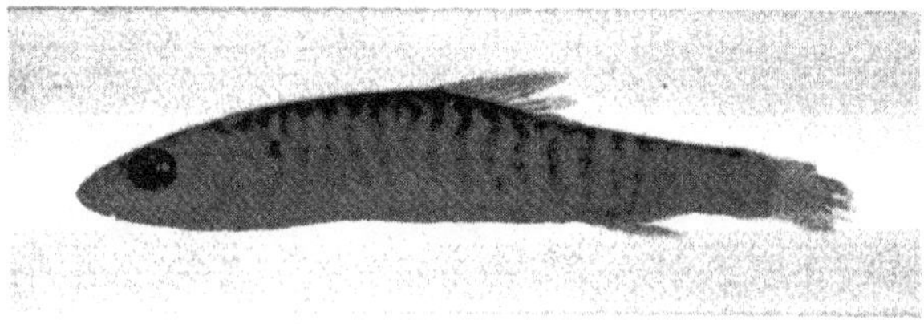

Fig. 23. Yunnanilus sp. Fuxian Lake, YU uncat., 41.9 mm SL.

lateral lines; *Y. macrogaster* has a smaller eye (3.3–3.5% SL versus 4.5) and *Y. parvus* a larger one (5.5–6.7); *Y. pleurotaenia* has a deeper head at nape (15.4–16.6% SL versus 13.8).

Yu Gu is a temporary lake (26°03′N 104°02′E) in Zhanyi Co. flowing to Bei Pan River, a tributary of Nanpan-jiang.

Yunnanilus sp. Fuxian Lake (Fig. 23)

Several specimens of this species are preserved in the Department of Biology, Yunnan University, Kunming. One of them, 41.9 mm was photographed by the first author who unfortunately had no time to examine it. It is strikingly different from any other species in the genus and may possibly represent the juvenile of a species in the *Y. nigromaculatus* group. Morphometric data obtained from the photograph are (as % SL): total length 115.6, dorsal head length 23.2, lateral head length 26.1, predorsal length 56.3, prepelvic length 59.0, preanal length 76.0, length of caudal peduncle 20.0, depth of caudal peduncle 10.0, snout length 8.2, eye diameter 8.2, body depth 19.3, head depth 15.5. Dorsal fin with $8^1/_2$ branched rays. The eye is larger than in any other species of *Yunnanilus* (8.2% SL versus 2.9–6.7) except *Y. brevis* which comes closer with 6.2–7.6. The body is more slender than in any other *Yunnanilus* without lateral line and cephalic pores (body depth 19.3% SL versus 20.9–28.2), the caudal peduncle longer (20.0% SL versus 8.9–14.4; 2.00 times longer than deep versus 0.68–1.35) and its depth overlaps lowest value of other *Yunnanilus* (10.0% SL versus 10.0–12.9), the head is longer (dorsal head length 23.2% SL versus 15.2–21.0; but this is difficult to measure with precision from a photograph) and preanal length is shorter (76% SL, 79.1–89.1).

Its habitat, Fuxian Lake, is described above under *Y. pleurotaenia.*

Yunnanilus sp. Mengzi Co. (Fig. 24)

Two specimens preserved in Kunming Institute of Zoology (KIZ 854381–382, 54.1–64.1 mm SL) may possibly represent a new species which is distinguished by the following combination of characters: incomplete lateral line with 14–17 pores; cephalic pores present; $9^1/_2$ branched dorsal rays; body depth 18–20% SL. Precise locality data are lacking as the specimens were received from local fishermen in Mengzi (23°23′N 103°23′E); this county being drained by both Red River and Nanpan-jiang (and several minor endorheic basins). Although *Yunnanilus* is not yet known from the Red River basin, we expect it to occur in that drainage too, but do not want to name another species without precise locality data.

Fig. 24. Yunnanilus sp. Mengzi Co., Yunnan, KIZ 854381, 64.1 mm SL.

Fig. 25. Yunnanilus sp. Er-Hai Lake, Yunnan, BMNH 1908.2.27:13–16, 65.1 and 58.9 mm SL.

Yunnanilus sp. Er-Hai (Fig. 25)

The first author examined 3 specimens (BMNH 1908.2.27: 13–16), 54.8–65.1 mm SL, identified as *Y. pleurotaenia* from Tongchuan-fu, a locality at the northern extremity of Er-Hai (= Tali Lake [also spelt Dali]). Hsiao (1949) and Ley et al. (1963) listed this species from Er-Hai and Zhu & Wang (1985) included Er-Hai in the distribution of the species. These three specimens are in a very poor state but apparently differ from the material we identify as *Y. pleurotaenia* by the following characters: caudal peduncle shallower (depth 8.0% SL versus 9.7–11.3; 39–43% HL versus 49–62; 1.7 times in length versus 1.2–1.4), snout shorter (8.1–8.8% SL versus 8.8–10.2; in % HL, its values overlap those of other species: 43–44 versus 42–52); narrower head width (10.3–11.7% SL versus 13.1–14.1; 56–57% HL versus 65–73); processus dentiformis very weak; 8 (versus 11) preoperculo-mandibular pores on cephalic lateral line system. They also seem to grow larger than any *Y. pleurotaenia* known to us (65.1 mm SL versus 56.7), but this might be an eco-phenotypic effect.

Berg (1938) received a specimen from the same series, a ripe female about 66 mm SL (now ZIL 25429) and illustrated it. Measurements from the drawing are intermediate between those of the specimens examined and our material of *Y. pleurotaenia*. Some of these differences are possibly due to the poor state of preservation of the specimens; thus new collections would be of the greatest interest and will show if the Er-Hai material represents a distinct species.

Er-Hai is situated in Dali (or Tali) Valley at an altitude of 1974 m; it is oriented North-South and is bordered at the west by the steep slopes of the Cang Shan mountains (4122 m). Its greatest length is about 40.5 km while its greatest width is 8.6 km and the average width about 6 km. Maximum depth is 20 m and average depth about 12 m. Water temperature varies between 22.4–25.0° C and monthly average air temperatures at Dali range between 9.8–20.8° C. The pH is between 7.9–8.5 (data from Ley et al. 1963, Hsiao 1946, 1949; see also Fitzgerald 1942 for data on topography). The basin of present Er-Hai (which now belongs to the Mekong drainage) was most probably part of a paleo-Red River which was continued upstream by the present day Jinsha-jiang (Yangtze) north of Lijiang (Gregory 1925, also Gregory & Gregory 1923). Er-Hai itself formed in the early Pleistocene (Yang 1984).

According to Li (1982), Er-Hai contains 17 species of fishes, 9 of which are endemics.

Water from Er-Hai leave the basin on its south-western corner, and after about 20 km joins the Yangbi-jiang. Effluent waters of Er-Hai are among the most polluted that the first author has seen; in

June 1986, at places its surface was packed by thick layers (up to one meter) of thick and hard foam. A short time spent collecting a few kilometres downstream of its confluence with Yangbi-jiang (a quite clear and clean stream) did not yield any fish.

Discussion

As presently known, the genus *Yunnanilus* is restricted to the Yunnan plateau of China and Inlé Lake, Burma. Future collections may allow detailed descriptions of the unnamed species listed above and will undoubtedly extend the range of the genus. However, it seems worth mentioning some important points in the distribution and evolution of the genus.

The relationships of *Yunnanilus* to other nemacheilines are still unresolved. Berg (1938) by naming the taxon *Eonemachilus* possibly considered it as primitive nemacheiline, but he did not discuss its relationships. Sawada (1982) discussed the relationships of *Y. pleurotaenia* to some nemacheilines, and found that *Y. pleurotaenia* was most closely related to *Orthrias* and *Triplophysa postventralis*. However, his study was somewhat inconclusive about interrelationships among nemacheilines and in any case was based on too few species of this subfamily to be accepted without further research; in addition it was based on osteological characters only and thus unique synapomorphies (like the two widely separated nares of *Yunnanilus* or the peculiar sexual dimorphism of *Triplophysa*) were not taken into consideration.

We did not investigate relationships among *Yunnanilus* for we think that this would be better left until more material is available and the remaining species are described. It is tempting to consider those species with 14 branched caudal rays and without a lateral line and cephalic pores as a distinct unit (for example at the sub-genus level, for which the name *Eonemachilus* would be available) but this would have to rely on degenerative characters which may have evolved independently several times. But, as noted by Johnson (1980, p. 42) '... the sharing of a number of reductive specializations in identical states is a reasonable indicator of affinity, even if each character cannot be shown to be uniquely derived'.

It seems obvious that *Y. nigromaculatus, Y.* sp. 'Yangtzung Lake' and *Y.* sp.cf. *nigromaculatus* from Er-Hai (and possibly *Y.* sp. 'Fuxian Lake') form a distinct group characterized by the anterior mouth, strong notched lower jaw with upwardly folded margins (Fig. 4) and the striking colour pattern; *Yunnanilus niger* is probably related to them too. *Yunnanilus paludosus* is possibly most closely related to *Y.* sp. 'Yu Gu' and both have resemblance to *Y. macrogaster.* Affinities of other species are difficult to hypothesize.

A striking trait in *Yunnanilus* is that most species occur in peculiar habitats which owe their peculiarity to karstic systems. Inlé Lake in Burma is a remnant of a former more extensive lake in a karstic area. The eastern Yunnan Plateau is part of an extensive karstic area also covering the whole Guizhou Province and west Guangxi and Viet Nam north of the Song Lo river. This huge karst area can be divided in three main types: the peak-forest type in the east, the plateau-canyon type in the center and the block-lake basin type in the west [the reader is referred to Zhang (1980) for a detailed description of karst in China]. The last two categories occur in the distribution area of *Yunnanilus.* The plateau-canyon type consists of shallow, gentle depressions and hills with 'senile' hydrographic networks (gentle slopes, slow current) in the upper reaches of rivers; downstream of the knickpoint, the wide valleys become deeply incised canyons with a much greater gradient. The area of Luoping (see picture 18 in Zhang 1980), Zhanyi and Xuanwei belongs to the plateau landscape of the plateau-canyon type. The block-lake type is characterized by differential movement of fault blocks which produced two kinds of landscape: block mountains and lake basins; these lakes are Dianchi Lake, Fuxian Lake, Yangtzung Lake, etc. Dali Lake does not belong to this group, but also formed in the early Pleistocene (Yang 1984).

The fish fauna of these habitats is not very diverse but includes a great number of endemics (68% of the lake fishes are endemics according to Li 1982; although his nomenclature is not very up to date, this figure seems to approximate the reality).

Endemics occur in several groups, but particularly remarkable are the genera *Anabarilius* and *Cyprinus*. Both speciated extensively in the lakes of the Yunnan Plateau [see Chen (1987) for a revision of *Anabarillius* and Zhou & Chu (1986) for a revision of Chinese *Cyprinus*]. If *Anabarilius* is restricted to the Nanpan-jiang and Jinsha-jiang (Yangtze) drainages, *Cyprinus* also occurs in the Red River, Mekong and Salween drainages; *Cyprinus* even evolved an endemic species in Inlé Lake (*C. intha* Annandale, 1918, originally described as a subspecies of *C. carpio* Linnaeus, 1758 but easily distinguished from any other species in the genus by its lower lateral line scale and branched dorsal fin ray counts); thus the distribution of *Cyprinus* in Yunnan and Burma parallels that of *Yunnanilus*. Also worth mention is *Sinocyclocheilus* which is always associated with caves (Chu & Cui 1985) and is endemic to Yunnan and western Guizhou.

It must also be mentioned that no other nemacheilines have been collected in any of the known *Yunnanilus* localities (with the exception of Inlé Lake); reports of *Triplophysa grahami* (Regan, 1906) (as a *Nemacheilus*) and '*Paracobitis*' *variegata* (Sauvage & Dabry, 1874) (as *N. oxygnathus* Regan, 1908) in Dianchi Lake are unlikely and most probably refer to inlets or outlets of the lake.

It is very tempting to put forward an hypothesis to explain *Yunnanilus* (and maybe *Cyprinus*) distribution and speciation on the Yunnan Plateau, namely, that *Yunnanilus* formerly inhabited an extensive Yunnan Plateau drainage which was broken up into isolated drainages. This seems supported by the wide and disjunct distribution of the specialized shrimp-eaters (*Y. nigromaculatus*, *Y.* sp. cf. *nigromaculatus* from Er-Hai, *Y.* sp. 'Yangtzung Lake' and probably *Y. niger* and *Y.* sp. 'Fuxian Lake' who live in mid-waters and which seem unable to survive in environments other than lakes or swamps. These smaller drainages, while providing the isolation which is a pre-requisite for allopatric speciation also acted as barriers against dispersal and invasion of potential competitors. The mechanisms which would explain the breakdown of a former extensive system are of two kinds: (a) normal evolution of the hydrographic system in a limestone area from a regular surface system to a system including sinkholes, underground rivers, doline lakes, etc., (b) superimposition of a new hydrographic system with high gradient streams (meaning a strong current and headwards erosion) which would either capture or isolate parts of the former system. The karst hypothesis certainly explains the isolation of most Yunnanese species, while the second hypothesis would explain the distribution of *Yunnanilus* and *Cyprinus* in both the Yunnan Plateau and Inlé Lake (which, with 47% endemism among its fish species is somewhat similar to the Yunnan lakes). Inlé Lake is separated from the Yunnan plateau by the high gradient Salween (or Nu-jiang) which flows in a deep canyon (up to 1500 m deep) and by the Mekong (or Lancang-jiang) which has a part of its course parallel to the Salween in its own canyon. This could imply that *Yunnanilus* was already present before or at the beginning of Himalayan orogeny; according to Gregory (1925) the main rivers of eastern Asia appeared at Oligocene and their present course dates back to Pliocene (Gregory & Gregory 1923).

Alternative or complementary hypotheses certainly exist and we envisaged several. The one we present above seemed to be the most likely. It can be tested by comparison with future geological researches (or researches unknown to us), investigation of relationships among *Yunnanilus* species and comparison with similar researches on other groups of animals with similar distribution patterns (including but not restricted to *Cyprinus*), and on investigations into the relationships of other endemics (like *Sawbwa*, *Sinocyclocheilus*, etc.).

As some trophic zones were not occupied, some *Yunnanilus* species could evolve specializations not met in any other member of the Cobitoidei permitting them to occupy zones usually occupied by members of other fish families (e.g. the anterior mouth with upwardly directed cleft of *Y. nigromaculatus* and *Y.* sp. Yantzong Lake; the [still to be confirmed] surface feeding of *Y. paludosus*). This kind of specialization has been called supralimital specialization by Myers (1960) who saw in species with such specializations the possible founders of new genera and higher categories. Myers also noted that most instances of supralimital specializations occured in lakes; although *Yunnanilus* also occur in streams, all those species which show specialization (supralimital or not) occur in lakes, marshes or small endorheic basins. Owing to these specializations they have been unable to reinvade streams, particularly those with a strong current, and this explains their limited distribution in south east Asia. After this manuscript was completed, Zhu Song-quan informed us that he is describing a remarkable new nemacheiline genus from Dianchi Lake; its peculiarities will possibly support our ideas discussed above.

Tatantze species-flock

The most remarkable point in the evolution and distribution of *Yunnanilus* is the presence of three sympatric species in Tatantze of Luoping Co., where only two other fish species are known (*Sinocyclocheilus malacopterus* and a unidentifiable *Discogobio*). A few other species have been introduced by riparians. Two hypothesis may explain their presence there: (1) all five native species may be remnants of the fauna of the stream in whose dry bed Tatantze marsh is located; (2) the marsh formed after the stream dried completely and the whole stream fauna disappeared, the marsh being colonized subsequently.

The first hypothesis does not fit well with the complete absence of other riverine fishes common in the area like *Misgurnus anguillicaudatus, Monopterus albus, Carassius auratus, Channa asiatica,* etc.; several of these species are air-breathers and would be expected in such habitats. If the second hypothesis is correct, we see two ways by which recolonization could be expected, viz. by surface waters (temporary flow, which is known to occur exceptionally after heavy rains) or underground waters. Both processes could also have occured. A strong point in favour of underground waters is that the marsh is still fed by underground water and that *Sinocyclocheilus, Garra* and its relatives (*Garra* is a genus to which *Discogobio* has very close affinities) and nemacheilines are known to invade caves and underground systems:

- *Sinocyclocheilus,* a genus endemic to the Yunnan Plateau, has been collected in caves (a new blind species from Iliang Co [25°24′N 103°27′E] is described by Chen & Chu, unpublished);
- *Garra* has been collected in caves in Oman (Banister 1984) and two closely related blind cave genera and species are known from Iraq (*Typhlogarra widdowsoni* Trewavas, 1955) and Iran (*Iranocypris typhlops* Bruun & Kaiser, 1950);
- *Yunnanilus parvus* is described here from the outlet of a cave and several species of nemacheilines have been described from caves, the species either:
  - with normal eyes and colour pattern [*Nemacheilus fasciatus* (Valenciennes, 1846) cited from the cave Guva Gremeng in Java by Weber & de Beaufort (1916) and Kottelat (1984)],
  - without colour but with normal eyes [*N. evezardi* Day, 1878 from an unspecified cave in India (Greenwood 1976) and a species from Siju Cave, India, identified as *Nemacheilus* sp. by Hora (1935), *N. beavani* Günther, 1868 by Hora (1935) and *N. multifasciatus* Day, 1878 by Pillai & Yazdani (1977)],
  - without colour and with degenerated eyes (an undescribed species from Thailand, Kottelat, unpublished),
  - without colour and eyeless [*N. smithi* Greenwood, 1976 from Iran; *N. starostini* Parin, 1983, from Turkmenia, *N. gejiuensis* Chu & Chen, 1979 from Gejiu Co. in Nanpan-jiang basin of Yunnan; *N. anophthalmus* (Zheng, 1981) from Yu-jiang basin in Guangxi (Fisheries Research Institute etc. 1981; Zhao 1983); *N. xiangxiensis* Yang, Yuan & Liao, 1986 from a tributary of Chang-jiang; an undescribed species from Thailand].

Due to the absence of other fishes, *Yunnanilus* species have been able to occupy various trophic zones not usually occupied by nemacheilines. While *Y. macrogaster* retained the usual nemacheiline feeding habit (but developed a stomach unique in the genus whose species otherwise have a nearly undifferentiated stomach), *Y. paludosus* has possibly developed surface feeding habits. The feeding habits of *Y. niger* are not known but its strong jaws with a well developed processus dentiformis on the upper jaw and a notch in the lower jaw seem to indicate a predator possibly paralleling *Y. nigromaculatus;* its body shape might indicate that it swims somewhat above the bottom. We suggest that it might be a shrimp or snail eater. Its colour pattern is also distinctive but we have no explanation to offer for this feature.

If *Y. niger* seems related to other *Yunnanilus* species without a lateral line and cephalic pores, the relationships of *Y. macrogaster* and *Y. paludosus* are not yet established; *Y. paludosus* shows some superficial resemblance to *Y.* sp. 'Yu Gu' of which we presently only have a photograph. The species we actually examined seem to be each other's closest relatives. If this is really the case, then the next question of interest, which future work with living specimens and future collections in drainages around Tatantze marsh may possibly solve, is to determine if they evolved sympatrically in situ, allopatrically in a subdivided pre-Tatantze, or resulted from multiple colonizations.

In refering to the Tatantze *Yunnanilus* species as a species-flock, we adopt the concept of Ribbink (1984), who consider that it is unnecessary for the flock to be monophyletic in the cladistic sense (as opposed to the definition of, e.g., Greenwood 1984). Like Ribbink, we think that the species flock concept draws attention to the high number of related species assembled in a narrowly defined area relative to the numbers of species in surrounding areas and not to the phylogeny of its various members. If the flock has to be monophyletic, then it is merely a (redundant) taxonomic category and it emphasizes its (hypothetical) phylogeny to the detriment of other biological information (e.g. disproportionate speciosity, endemicity).

Regarding the uncertainties surrounding the evolution and interrelationships among the Tatantze *Yunnanilus* species, it is possibly not strictly correct to refer to them as a species flock. Nevertheless, it is a convenient term as all species concerned are quite closely related and it denotes some similarities to certain of the species flocks already described, particularly those comprising few species, like the *Cyprinodon* species flock of Laguna Chichancanab, Yucatan (Humphries & Miller 1981, Humphries 1984).

Extensive new collections from the innumerable small endorheic basins and caves of the karstic area of Yunnan, Guangxi and Guizhou will certainly result in the discovery of additional species showing parallel and other specializations, and the discovery of similar species flocks. We also expect that cave-dwelling *Yunnanilus* species may be discovered.

Acknowledgements

We are particularly indebted to Yang Junxing, Cui Guihua and Chen Yinrui for their invaluable help in providing and locating information on the specimens and the localities. Ho Jichang (Department of Biology, Yunnan University, Kunming), Wang Xiurong (Yunnan Fisheries Institute, Kunming) and Martine Desoutter (MNHN) allowed access to material in their institutions. Alwyne Wheeler, Gordon Howes (BMNH) and Gianna Arbocco (MCSNG) loaned material under their care. Volker Mahnert (MHNG) arranged some of the loans. Liang Xingcai identified insects in stomach contents. Figures 5, 15, 18 and 25 are by G. Dajoz (MHNG), 7, 9, 11 and 14 by Yang Junxing and Cui Guihua; other figures are by the first author. Most of this work was completed while the first author visited KIZ; this was made possible with financial support from Burrus S.A., Boncourt, Switzerland and Cercle d'Etudes Scientifiques, Société Jurassienne d'Emulation.

References

Annandale, N. 1918a. Introductory account of the Inlé Lake. Rec. Indian Mus. 14: 1–7, 1 pl.

Annandale, N. 1918b. Fish and fisheries of the Inlé Lake. Rec. Indian Mus. 14: 33–64, pl. 1–7.

Banister, K.E. 1984. A subterranean population of *Garra barreimiae* (Teleostei: Cyprinidae) from Oman, with comments on the concept of regressive evolution. J. Nat. Hist. 18: 927–938.

Bender, F. 1983. Geology of Burma. Borntraeger, Berlin. 293 pp.

Berg, L.S. 1938. On some South China loaches (Cobitidae, Pisces). Bull. Soc. Nat. Moscou, Ser. Biol. 47: 314–318 (in Russian and English).

Boulenger, G.A. 1893. List of fishes collected by Mr E.W.

Oates in the Southern Shan States, and presented by him to the British Museum. Ann. Mag. Nat. Hist. (6) 12: 198–203.

Bruun, A.F. & E.W. Kaiser. 1950. *Iranocypris typhlops* n.g., n. sp., the first true cave-fish from Asia. Danish Scient. Invest. Iran, Kobenhavn (1944[1950]) (4): 1–8.

Chaudhuri, B.L. 1911. Contributions to the fauna of Yunnan based on collections made by J. Coggin Brown. II. Fishes. Rec. Ind. Mus. 6: 13–23.

Chen Y.-R. 1986. A revision of fishes of the cyprinid genus *Anabarilius.* Acta Zootax. Sinica 11: 429–438 (In Chinese).

Chu X.-L. & Chen Y.-R. 1979. A new blind cobitid fish (Pisces, Cypriniformes) from subterranean waters in Yunnan, China. Acta Zoologica Sinica 25: 285–287 (In Chinese).

Chu X.-L. & Cui G.-H. 1985. A revision of Chinese cyprinid genus *Sinocyclocheilus* with reference to the interspecific relationships. Acta Zootax. Sinica 10: 435–441 (In Chinese).

Coggin Brown, J. 1910. Contributions to the fauna of Yunnan based on collections made by J. Coggin Brown, 1909–1910. Introduction. Rec. Indian Mus. 5: 193–196.

Fisheries Research Institute, Guangxi Zhuang Tribe Self-Governed Territory & Institute of Zoology, Academia Sinica. 1981. Freshwater fishes of Guangxi. Guangxi Renmin Press, Nanning. 257 pp.

FitzGerald, C.P. 1942. The Tali district of western Yunnan. Geogr. J. 99: 50–60.

Greenwood, P.H. 1976. A new and eyeless cobitid fish (Pisces, Cypriniformes) from the Zagros mountains, Iran. J. Zool. 180: 129–137.

Greenwood, P.H. 1984. What *is* a species flock. pp. 13–19. *In:* A.A. Echelle & I. Kornfield (ed.) Evolution of Fish Species Flocks, Univ. Maine at Orono Press, Orono.

Gregory, J.W. 1925. The evolution of the river system of south-eastern Asia. Scottish Geogr. Mag. 41: 129–141.

Gregory, J.W. & C.J. Gregory. 1923. The alps of Chinese Tibet and their geographical relations. Geogr. J. 61: 153–179, 4 pl.

Hora, S.L. 1924. Fish of the Siju Cave, Garo Hills, Assam. Rec. Indian Mus. 26: 27–31.

Hora, S.L. 1929. Notes on fishes in the Indian Museum. XVII. Loaches of the genus *Nemachilus* from Burma. Rec. Indian Mus. 31: 311–334, pl. 14–15.

Hora, S.L. 1935. Notes on fishes in the Indian Museum. XXIV. Loaches of the genus *Nemachilus* from eastern Himalayas, with the description of a new species from Burma and Siam. Rec. Indian Mus. 37: 49–67, 3 pl.

Hsiao S.C. 1946. A limnological study of Erh Hai, Yunnan, China: physico-chemical characteristics. J. Anim. Ecol. 15: 1–8.

Hsiao, S.C. 1949. A limnological study of Erh Hai. 2. Biological features. J. Anim. Ecol. 18: 89–99.

Humphries, J.M. 1984. *Cyprinodon verrecundus,* n. sp., a fifth species of pupfish from Laguna Chichancanab. Copeia 1984: 58–68.

Humphries, J.M. & R.R. Miller. 1981. A remarkable species flock of pupfishes, genus *Cyprinodon,* from Yucatán, México. Copeia 1981: 52–64.

Illick, H.J. 1956. A comparative study of the cephalic lateral-line system of North American Cyprinidae. Amer. Midl. Nat. 56: 204–223.

Jayaram, K.C. 1981. The fresh-water fishes of India, Pakistan, Bangladesh, Burma and Sri Lanka. A handbook. Zool. Surv. India, Calcutta. xxii + 475 pp.

Johnson, G.D. 1980. The limits and relationships of the Lutjanidae and associated families. Bull. Scripps Inst. Oceanogr. 24: 1–114.

Kottelat, M. 1984. Revision of the Indonesian and Malaysian loaches of the subfamily Noemacheilinae. Jap. J. Ichthyol. 31: 225–260.

Kottelat, M. 1986a. Die Fischfauna des Inlé-Sees in Burma. Aquar. Terrar. Ztschr. 39: 403–406.

Kottelat, M. 1986b. Die Fischfauna des Inlé-Sees in Burma (2). Aquar. Terrar. Ztschr. 39: 452–454.

Ley S.-H., Yu M.-K., Li K.-C., Tseng C.-M., Chen C.-Y., Kao P.-Y. & Huang F.-C. 1963. Limnological survey of the lakes of Yunnan plateau. Oceanologia et Limnologia Sinica 5: 87–114 (In Chinese).

Li H. 1980. A study on lake vegetation of Yunnan Plateau. Acta Bot. Yunnanica 2: 113–118 (In Chinese).

Li S.-S. 1982. Fish fauna and its differentiation in the upland lakes of Yunnan. Acta Zoologica Sinica 28: 169–176 (In Chinese).

Myers, G.S. 1960. The endemic fish fauna of Lake Lanao, and the evolution of higher taxonomic categories. Evolution 14: 323–333.

Nichols, J.T. 1925. *Nemacheilus* and related loaches in China. Amer. Mus. Novit. 171: 1–7.

Parin, N.V. 1983. *Noemacheilus (Troglocobitis) starostini* sp. n. (Osteichthyes, Cobitidae) a new blind fish from subterraneous waters of Kugitangtau (Turkmenia). Zool. Zh. 62: 83–89 (In Russian).

Pillai, R.S. & G.M. Yazdnani. 1977. Ichthyo-fauna of Garo Hills, Meghalaya (India). Rec. Zool. Surv. India 72: 1–22.

Regan, C.T. 1904. On a collection of fishes made by Mr. John Graham at Yunnan Fu. Ann. Mag. Nat. Hist. (7) 13: 190–194.

Rendahl, H. 1948. Die Süsswasserfische Birmas. I. Die Familie Cobitidae. Ark. Zool. 40A (7): 1–116.

Ribbink, A.J. 1984. Is the species flock concept tenable? pp. 21–25. *In:* A.A. Echelle & I. Kornfield (ed) Evolution of Fish Species Flocks, Univ. of Maine at Orono Press, Orono.

Sawada, Y. 1982. Phylogeny and zoogeography of the superfamily Cobitoidea (Cyprinoidei, Cypriniformes). Mem. Fac. Fish. Hokkaido Univ. 28: 65–223.

Sun X., Wu Y., Qiao Y., & D. Walker. 1986. Late Pleistocene and Holocene vegetation history at Kunming, Yunnan Province, southwest China. J. Biogeogr. 13: 441-476.

Tchang S. 1948. Recherches limnologiques et zoologiques sur le lac de Kunming, Yunnan. Contr. Inst. Zool. Natn. Acad. Peiping 4: 1–24.

Trewavas, E. 1955. A blind fish from Iraq, related to *Garra.* Ann. Mag. Nat. Hist. (12) 8: 551–555.

Weber, M. & L.F. de Beaufort. 1916. The fishes of the Indo-Australian Archipelago. 3. Brill, Leiden.

Yang G.-Y., Yuan F.-X., & Liao Y.-M. 1986. A new blind

Cobitidae fish from the subteranean water in Xiangxi, China. J. Huazhong Agric. Univ. 5 (3): 219–223 (In Chinese).
Yang L.F. 1984. The preliminary study on the original classification and distributional law of lakes on the Yunnan plateau. Trans. Ocean. Limnol. 1: 34–39 (In Chinese).
Zhang Z.-G. 1980. Karst types in China. GeoJournal 4: 541–570.
Zhao Z.-R. 1983. New discovery of a cave blind fish in Wuming, Guangxi. Carsologica Sinica 1983 (1): 57–80 (In Chinese).
Zhou W. & Chu X.-L. 1986. Systematic study of the genus *Cyprinus* (Pisces: Cyprinidae) in Yunnan, China. Zool. Res. 7: 297–310 (In Chinese).
Zhu S.-Q. 1981. Notes on the scaleless loaches (Nemachilinae, Cobitidae) from Qinghai-Xizang Plateau and adjacent territories in China. Geological and Ecological Studies of Qinghai-Xizang Plateau 2: 1061–1070.
Zhu S.-Q. 1986. A comparative study on the air-bladder and its bony capsule of noemacheiline fishes (Cobitidae) in China. Acta Hydrobiol. Sinica 10: 136–143 (In Chinese).
Zhu S.-Q. & Wang S.-H. 1985. The noemacheiline fishes from Yunnan Province, China (Cypriniformes: Cobitidae). Acta Zootax. Sinica 10: 208–220 (In Chinese).

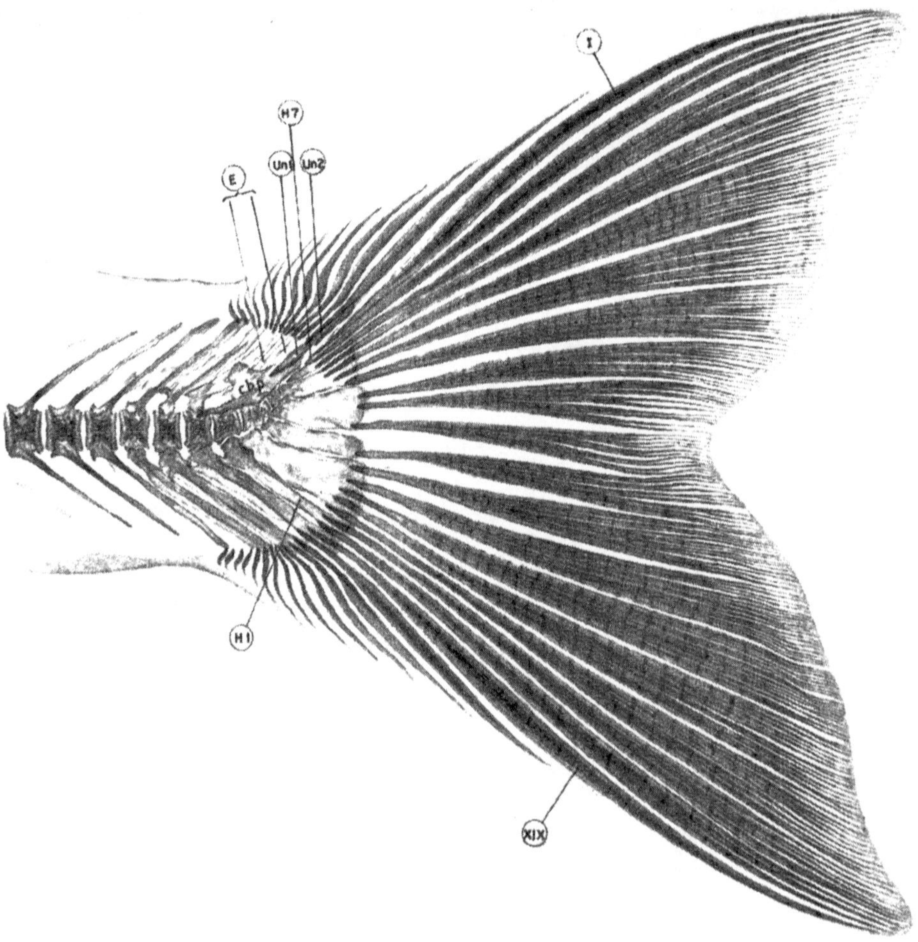

Caudal skeleton of landlocked sockeye salmon from New York by Paul I. Voevodine for Vladykov (1962).

Environmental Biology of Fishes Vol. 23, No. 1–2, pp. 95–113, 1988

Taxonomic status of the extinct Banff longnose dace, *Rhinichthys cataractae smithi*, of Banff National Park, Alberta

Claude B. Renaud[1] & Don E. McAllister[2]
[1] *Département de Biologie, Université d'Ottawa, Ottawa, Ontario K1N 6N5, Canada*
[2] *Ichthyology Section, National Museum of Natural Sciences, P.O. Box 3443, Ottawa, Ontario K1P 6P4, Canada*

Received 25.11.1987 Accepted 12.12.1987

Key words: Endangered species, Banff National Park, Cave and Basin Hotsprings, Introductions, Exotic species, Cyprinids

Synopsis

Study of 314 specimens of *Rhinichthys cataractae* from British Columbia, Alberta, and Wyoming, lead to the following conclusions: (1) *Rhinichthys cataractae smithi* Nichols, 1916, is a valid subspecies, endemic to Cave and Basin Hotsprings and distinguished by 48–58 as opposed to 58–74 lateral line scales; (2) between 1925 and 1971, *R. c. smithi* hybridized with the eastern subspecies *R. c. cataractae* (Valenciennes, 1842) from the Bow River and by 1981 the former had undergone almost complete introgression and was virtually extinct; (3) probable factors leading to this are introduction of tropical fishes into the hotsprings and periodic reduction of inflow from the hotsprings; (4) the closest relative of *R. c. smithi* is *R. c. cataractae,* rather than the westslope longnose dace (without a scientific name) inhabiting the Pacific basin; (5) the low number of lateral line scales of *R. c. smithi* may be a pleomeristic response to dwarfing; (6) *R. c. smithi* develops breeding tubercles at sizes as small as 21.1 mm SL, whereas *R. c. cataractae* develop them at 36.3 mm SL in Alberta; (7) introductions should not be made into a body of water prior to the study of its native fishes and consultation with experts in taxonomy and distribution of rare fishes. *R. c. smithi* is illustrated for the first time.

Introduction

Longnose dace have been known to occur in the hotsprings at Banff, Banff National Park, Alberta, Canada for at least 90 years. First collected in 1892 (Eigenmann 1894), they were described by Nichols (1916) as a distinct subspecies, *Rhinichthys cataractae smithi* (Fig. 1). Because the subspecies was endemic to the hotsprings, because the environment underwent profound changes which endangered the population, and because subspecies concepts have changed since 1916 it was decided to re-examine the taxonomic status of this subspecies.

In this paper we attempt to answer several questions: (1) whether the original Cave and Basin Hotspring dace population is still surviving or has been replaced by another, (2) whether *Rhinichthys cataractae smithi* Nichols, 1916 is a valid subspecies, taxonomically distinct from neighbouring populations, (3) if it is a valid subspecies, are its relationships closer to eastern (Arctic, Hudson Bay and Gulf of Mexico, and Atlantic drainages) or to Pacific basin populations? (4) what is the taxonomic status of neighbouring eastern and Pacific basin populations.

Our study does show the Banff longnose dace to

Fig. 1. Photograph of the holotype (AMNH 5514) of the Banff longnose dace, *Rhinichthys cataractae smithi* Nichols, 1916. The standard length is 36.4 mm. Photographed by Carl Ferraris, AMNH.

have been a distinct subspecies. Since very little had been previously known about this subspecies and since it had become extinct we thought it provident to record much of what we knew while it was still available. Because of the lack of information we have had to contribute towards a better taxonomic understanding of the *R. cataractae* complex in order to determine the relationships of the Banff longnose dace.

The Cave and Basin Hotsprings (51° 10′N, 115° 35′ 20″W), are located about 2 km southwest of the centre of the townsite of Banff and are tributary to the Bow River in the South Saskatchewan River Basin (Fig. 2). Basin Spring flows north into the southwest corner of Cave and Basin Hotspring Pond, Cave Spring north into the middle of the south shore of the pond, and Pool Spring into the east end of the pond. Inhabitable parts of the spring streams are limited to a few metres. The pond, shallow and marshy, is about a third of a km long and at an altitude of 1382 m. It has two outlets on the north side to the Bow River, an eastern one via a small surface flow over a beaver dam and a western one, halfway along the northern shore, blocked by an earthfill almost 1 m higher than the river level. Aerial photos taken from 25 July 1950 to 12 May 1983 show little difference in size of the pond. More information is given on the Cave and Basin Hotsprings drainage by Nelson (1983).

History

The hotsprings were known to Stoney and other indians of the region but were only 'discovered' by the Palliser Expedition in 1859. In 1885 the government set up a small area around the hotsprings as the first national park in Canada. In 1887 a wooden bath-house was built at the springs and in 1914 a grandiose swimming pool with stone arches and inlaid floors, which was supplied by the springs, was opened (preceding history largely drawn from Smyth 1985). Closed completely in 1976, the public baths were renovated and re-opened in 1985 as part of the national park's centennial celebrations.

Dace were first recorded in the hotsprings in 1892 by Eigenmann (1894) who noted them as being very abundant. He reported them under the name *Rhinichthys dulcis* which is currently regarded as a synonym of *Rhinichthys cataractae* or as a subspecies of that species. There is a prior unpublished report by McCardell (see Appendix), dated November 1883, who claimed to have seen a school of little minnows in the stream (no longer extant) which flowed along the bench for about 30 m and then tumbled down the bench into the lower circular basin. This is interesting because collections made in the 1970s and 1980s were all below the lowest bench in hotspring outflows slightly above the level of the Bow River. Nichols (1916) named specimens collected from the hotspring at Cave and Basin by Harlan I. Smith of the Victoria Memorial Museum (former name for National Museums of Canada) as a new subspecies, *Rhinichthys cataractae smithi*, distinguishing it from *R. cataractae dulcis* by the lower lateral line scale count, 50 to 60 scales.

Since its scientific description, the Banff longnose dace has received relatively little notice by scientists and an illustration has never been published. Jordan et al. (1930) even omitted it from their list of fishes of North America, Dymond (1947) from his list of fishes east of the Rockies, Slastenenko (1958), and Scott & Crossman (1973) from their *Freshwater fishes of Canada*. However, Paetz & Nelson (1970) discussed *R. c. smithi* and noted that R.B. Miller had specimens collected from the hotsprings on 27 July 1924 and had re-

Fig. 2. Aerial photograph of the Cave and Basin Hotsprings taken 12 May 1983 from an altitude of 2286 metres (National Air Photo Library roll no. 26217).

ferred them to *Agosia nubilis* (a synonym of a subspecies of *R. osculus*). Nelson & Paetz (1982) noted that the taxonomic status of the nominal subspecies, *R. c. smithi,* was under study and that it was unknown whether the surviving hotspring population represented the original form. Nelson (1983) and Crossman (1984) reported on introductions into the hotsprings. Campbell (1985) listed *R. c. smithi* with a provisional status of endangered.

McAllister (1969) noted that *R. c. smithi* was not represented in a collection, comprising introduced tropical fishes, taken from the hotspring in June 1968 and suggested (1970) it was possibly extinct due to the introduction of the tropical fishes. McAllister & Gruchy (1977) remarked that it was still surviving in 1971 but endangered by introduction of tropical aquarium fishes. Lanteigne (1981) sampled the hotsprings, evaluated the status of the

Banff longnose dace, and recommended to COSEWIC that it be given endangered status due to the threat of introduced fishes and the alteration of habiat. Nelson (1983) discussed the history of the introductions into the hotspring and reported 8 specimens of longnose dace taken in 1981 but added that it was unknown whether the existing specimens were descendant from the population which existed before man-made changes to the system. He also reported a personal communication indicating a belief in movement of dace in recent years between the hotspring and the Bow River (directions not stated). Nelson (1983) raised the question whether, if such a migration takes place, the original population made such a migration or whether it died out and was replaced by Bow River stock. Anonymous (1987) reported that Parks Canada accidentally almost drained the pond below the Cave and Basin Hotsprings in 1987. The beaver dam impounding the pond broke when park wardens attempted to drain water through the dam. A nearby horsetrail had been flooded and draining was attempted in order to dry the trail. No information is available on whether the hybrid longnose dace or introduced tropical fishes survived the draining. It is reported (Dave Mayhood, unpublished) that wardens blocked the outflow and refilled the pond as quickly as they could, but not before a substantial portion of the pond bottom had been exposed.

There have been few taxonomic studies of the *Rhinichthys cataractae* and related species complexes in recent years. Butcher (1980) studied hybridization between *Rhinichthys cataractae* and *Couesius plumbeus* in the Kananaskis reservoirs, Alberta. McPhail (1967, 1980) and Bisson & Reimers (1977) studied geographic variation of *Rhinichthys cataractae* populations in Oregon, Washington and Idaho. Stauffer et al. (1979) and Morgan et al (1984) studied *Rhinichthys bowersi,* a possible hybrid or species resulting from a *Nocomis micropogon* and a *Rhinichthys cataractae* cross. Miller (1984) described a new species, *Rhinichthys deaconi* from Nevada, in the *R. osculus* species complex.

Material

A total of 314 specimens of *Rhinichthys cataractae* from Alberta and British Columbia, Canada and Wyoming, United States and preserved in 50% isopropanol or 70% ethanol were studied. The material studied is listed further in order of date of collection within four geographic groupings. Figure 3 shows the locations of the collection sites.

The following abbreviations are used for museum collections: AMNH – American Museum of Natural History, New York; IU – Indiana University (now at University of Michigan Museum of Zoology, Ann Arbor); NMC – National Museum of Natural Sciences, National Museums of Canada, Ottawa; ROM – Royal Ontario Museum, Toronto; UAMZ – University of Alberta Museum of Zoology, Edmonton; UCMZ – University of Calgary Museum of Zoology, Calgary; UMMZ – University of Michigan Museum of Zoology, Ann Arbor.

Temperature data are included for the Cave and Basin collections whenever available.

From Banff Cave and Basin Hotsprings

128 specimens from 18.8 to 55.0 mm SL:

IU 4409, Alberta: hotsprings at Banff, collected by C.H. Eigenmann, August or September 1892, (8), 28.8–34.0 mm SL.

AMNH 5514, Alberta: hot spring at Cave and Basin, Banff, collected by H.I. Smith, 21 July 1915, holotype, (1), 36.4 mm SL.

AMNH 17368, Alberta: hot spring at Cave and Basin, Banff, collected by H.I. Smith, 21 July 1915, paratypes (4), 23.4–37.1 mm SL.

ROM 1713, Alberta, Sulphur Springs, Banff, collected by E.H. Craigie, 18 June 1925, (6), 21.25–28.7 mm SL, nuptial tubercles present on 3 of 3 examined.

NMC58-226, Alberta: hotsprings at Banff, Rocky Mountain region, (84), 18.8–46.8 mm, collection date unknown, certainly before 1958 and probably between 1920 and 1940, nuptial tubercles present on 61 of 64 examined.

NMC71-218, Alberta: Cave and Basin Hotsprings, outlet creek, Banff National Park, collected by J.C. Ward, 14 May 1971, (16), 41.3–55.0 mm SL, 26° C, 20 cm water depth, nuptial tubercles present on 16 of 16.

UAMZ 4613, Alberta: cave at east branch of Cave Hotspring, Banff National Park, collected by J.S. Nelson et al., 16 May 1981, (1), 51.2 mm SL, 22.5–24° C, nuptial tubercles present.

UAMZ 4614, Alberta: southwest end of Cave and Basin drainage area, Banff National Park, collected by J.S. Nelson et

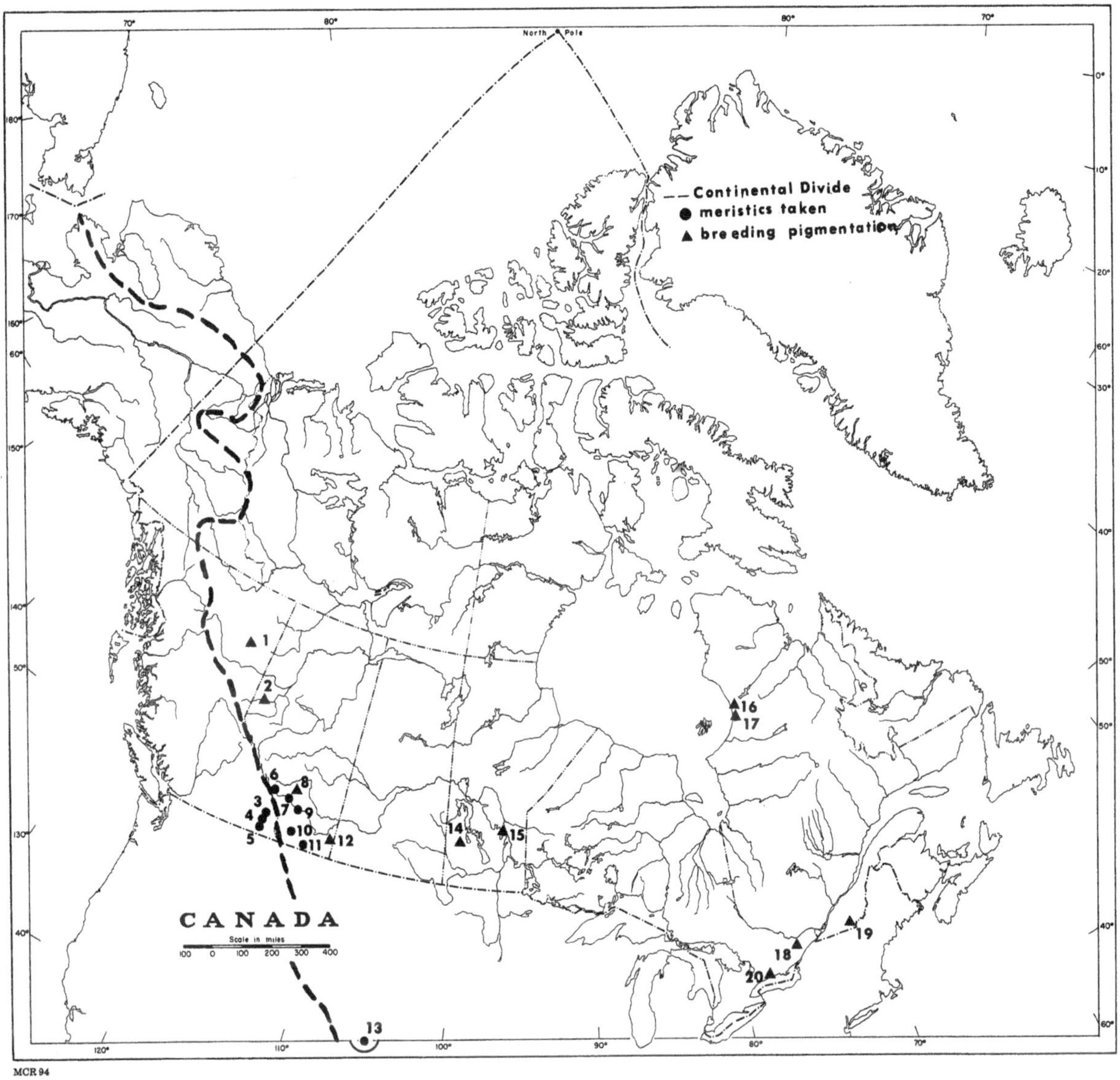

Fig. 3. Location of study samples. 1– NMC81-778, 2– NMC70-114, 3– NMC 66-68, 4– NMC76-71, 5– NMC 67-280, 6– Cave & Basin Hotsprings (see Material), 7– NMC 64-877, 8– Red Deer River basin (Paetz & Nelson 1970), 9– UMMZ 1977– 22353– 22487, 10– NMC67-694, 11– NMC66-298, 12– Cypress Hills (Paetz & Nelson 1970), 13– UMMZ 127544, 14– Mink & Valley rivers (Bartnik 1972), 15– Lake Winnipeg (Keleher 1952 in Scott & Crossman 1973), 16– NMC 63–213, 17– NMC63-221, 18– NMC 65-133, 19– NMC77-700, 703, 712, 713, 20– ROM 14260.

al., 17 May 1981, (1), 32.8 mm SL, 21° C, nuptial tubercles absent.

UAMZ 4615, Alberta: cave at east branch of Cave Hotspring, Banff, collected by J.S. Nelson et al., 17 May 1981, (5), 45.5–53.2 mm SL, 22.5–24° C, most orange coloured in part of snout and mouth area and assumed to be males in breeding condition, nuptial tubercles present on 5 of 5.

NMC81-1159, Alberta: Cave and Basin marsh, above bridge, along foot path, Banff National Park, collected by J. Lanteigne and F. McAllister, 16 September 1981, (1), 27.0 mm SL, 17° C, water 10 cm to 2 m deep, nuptial tubercles absent.

NMC81-1160, Alberta: pool above beaver dam, Cave and Basin hotsprings area, Banff National Park, collected by J. Lanteigne and F. McAllister, 17 September 1981, (1), 27.5 mm SL, 17° C, nuptial tubercles absent.

From Bow River, Alberta

66 specimens from 17.0 to 89.5 mm SL from the Bow River.

ROM 12111, Alberta: Bow River, August 1938, (3), 17.0–46.2 mm SL, nuptial tubercles present on 1 of 3.

ROM 25921, Alberta: Bow River, (35), 45.5–89.5 mm SL, nuptial tubercles present on 25 of 35.

UCMZ 1977-22353-22487, Alberta: Bow River below Carlsland Dam, 31-21-25-W4, collected by J. Elliot, 24 July 1965, (28), 31.5–51.4 mm SL, nuptial tubercles present on 3 of 4 examined.

From the Rest of Southern Alberta

56 specimens from 25.6 to 56.6 mm SL from southern Alberta other than Cave and Basin Hotsprings and the Bow River.

NMC64-877, Alberta: 8.1 km NW of Priddis, collected by F.R. Cook and C.B. Powell, 13 June 1963, (7), 25.6–29.3 mm SL.

NMC66-298, Alberta: Milk River, R15, TP1, S35, NE 1/4, 12.9 km east and 6.5 km south of town of Milk River, collected by T.A. Willock, 18 May 1966, (37), 25.8–56.6 mm SL, nuptial tubercles present on 30 out of 30 examined.

NMC67-694, Alberta: Oldman River Tp11, R16, S6, SW 1/4, 10 km north and 1.6 km west of Taber, collected by T.A. Willock, 23 August 1963, (12), 30.8–46.2 mm SL.

British Columbia

41 specimens from 40.2 to 86.9 mm SL.

NMC76-71, British Columbia: Skookumchuck, Kootenay River, east side, collected by T.J.M. Webster, 15 August 1965, (9), 40.2–86.9 mm SL.

NMC66-68, British Columbia: Kootenay River, Canal Flats, collected by T.J.M. Webster, 16 August 1965, (30), 41.2–66.2 mm SL, nuptial tubercles present on 17 of 30.

MNC67-280, British Columbia: 14.5 km SW of Cranbrook on Highway 3, collected by F.R. and J.C. Cook, 12 June 1967, (2), 44.0–44.6 mm SL.

Wyoming

25 specimens from 33.7 to 66.0 mm SL.

UMMZ 127544, Wyoming: Sweetwater River at Hudson's Ranch at crossing of U.S. 87 near Casper, Fremont Co., C.L. and L.C. Hubbs, 11 July 1934, (25), 33.7–66.0 mm SL.

Methods

Specimen measurements using dial calipers were recorded to the nearest 0.05 mm. Except as noted, the measurements follow Hubbs & Lagler (1964): standard length, snout length, head length (excluding opercular membrane), opercular membrane length (taken at the level of the middle of the orbit), frenum width (taken transversely), edge of maxilla to snout, posterior end of frenum to end of snout, least fleshy interorbital width.

Meristic counts also follow Hubbs & Lagler (1964): dorsal fin rays, pored lateral line scales, abdominal vertebrae (up to where the first anal pterygiophore intersects a vertebra and including the four modified vertebrae of the Weberian apparatus), caudal vertebrae (including the urostyle as one vertebra). Vertebral counts were made from radiographs. Sex was determined by dissection. Presence or absence of a maxillary barbel, nuptial tubercles (breeding tubercles) and a pelvic axillary scale were also recorded.

Specimens were separated into four geographic groups, Cave and Basin Hotsprings; Bow River; the rest of Alberta (excluding Cave and Basin Hotsprings and Bow River); and British Columbia as close as possible to Cave and Basin Hotsprings but in the Pacific drainage, as well as more distant populations. The Sweetwater River, central Wyoming sample was included because that is the type locality of *Argyreus* (= *Rhinichthys*) *dulcis* Girard, 1857.

We chose the probability of $p<0.02$ as indicating statistical significance.

Results

Meristics

Meristic counts and standard lengths are summarized in Table 1 and lateral line scale count frequencies are shown in Fig. 4. Morphometric data are deposited in the files of the Ichthyology Section, National Museum of National Sciences, Ottawa. The hotspring samples are summarized under three periods, 1892–1925, (*early* – 3 collections), 1920–40? (*mid* – 1 collection), and 1971–81 (*late* – 6 collections).

Mean values for lateral line scale and dorsal fin ray counts increase through the hotspring time se-

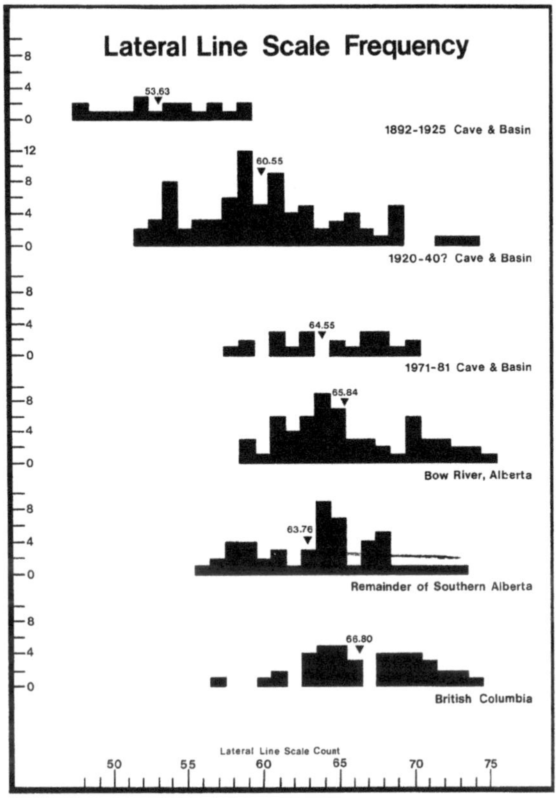

Fig. 4. Lateral line scale count frequencies in early, mid and late hotspring, Bow, rest of southern Alberta, and B.C. samples.

ries – early, mid and late (53.63, 60.55, 64.55 respectively for scales and 7.33, 7.63 and 7.92 for rays). The mean values and maximum values for standard length show a similar pattern. The presence of nuptial tubercles in some specimens from each sample shows that the small size of fish in the early sample cannot be attributed to their being immature young. The abdominal, caudal and total vertebral counts do not show a similar pattern; the mid sample is either higher or lower than the early and late samples. Vertebrae do not appear to be a good taxonomic character and we do not analyze them further in the hotspring samples. We found no sexual dimorphism in lateral line scale counts in specimens from the hotsprings. The trend for size, scale and ray counts to increase from early to late hotspring samples may suggest these meristic values were size dependent. We tested this relationship on the largest single Bow River sample (ROM 25921, 21 specimens) and all Bow samples (62 specimens). The correlations and probabilities between standard length and lateral line scales were respectively 0.16 and 0.12 and $p>0.49$ and >0.34. Dorsal rays were invariate, hence uncorrelated.

Chi square tests of lateral line scale counts (grouped into <59, 59–66, >66) between all pairs of early, mid and later samples were significant at a level of $p<0.002$. The early-late comparison was most significant at $p<0.00001$. Chi square tests of dorsal fin ray counts (grouped into 6–7 and 8) showed the early and mid samples were moderately different, $p<0.01$, and mid and late to be moderately different, $p<0.01$, while early and late samples were very different at $p<0.001$.

Correctness of identification percentages (percentages of specimens correctly identified above and below the most diagnostic value with 59 or fewer for scales and 7 or fewer for dorsal rays) show that the early and late samples are more distinct than the other and that 93.2% (i.e. $(19/19 + 19/22)/2 \times 100$) of the specimens were correctly allotted with lateral line scales and 79% with dorsal fin rays. Averaging both characters, the early-late samples were more distinct (86.0%), than early-mid (70.9%), or mid-late samples (65.3%).

It is also apparent from Table 1 that mean values for lateral line scales, dorsal fin rays and standard length in the Bow River sample are more similar to values in the late sample than to the early or mid samples. All values for the Bow River are slightly higher than those in the late sample (but not statistically indistinguishable). Overall the results show that the hotspring samples became progressively more like the sample from the Bow River, with the late sample being scarcely distinguishable from those in the Bow.

Male breeding color

Male breeding color bears on the taxonomy of the *Rhinichthys cataractae* complex, as we shall see further, and the paucity of color observations leads us to publish nuptial color from notes in NMNS records. McPhail & Lindsey (1970) and Bartnik (1972) described spawning color differences between populations east and west. Localities for the

Table 1. Frequency of meristic characters and SL in *Rhinichthys cataractae.*

	Pored lateral line scales														
	N	*X̄*	*SD*	*CV*	*48*	*49*	*50*	*51*	*52*	*53*	*54*	*55*	*56*	*57*	*58*
1892–1925 Cave & Basin	19	53.63	3.52	6.56	2	1	1	1	3	1	2	2	1	2	1
1920–40? Cave & Basin	82	60.55	5.09	8.41	–	–	–	–	2	3	8	2	3	3	6
1971–81 Cave & Basin	22	64.55	3.73	5.78	–	–	–	–	–	–	–	–	–	–	1
Bow River, Alberta	62	65.84	4.24	6.44	–	–	–	–	–	–	–	–	–	–	–
Rest southern Alberta	51	63.76	4.13	6.48	–	–	–	–	–	–	–	–	1	2	4
British Columbia	41	66.80	3.96	5.93	–	–	–	–	–	–	–	–	–	1	–
Wyoming	24	66.04	5.10	7.72	–	–	–	–	–	–	–	–	–	–	2

	Pored lateral line scales continued																	
		59	60	61	62	63	64	65	66	67	68	69	70	71	72	73	74	75
1892–1925 Cave & Basin	..	2	–	–	–	–	–	–	–	–	–	–	–	–	–	–	–	–
1920–40? Cave & Basin	..	12	5	9	4	5	2	3	4	2	1	5	–	–	1	1	1	–
1971–81 Cave & Basin	..	2	–	3	1	3	–	2	1	3	3	1	2	–	–	–	–	–
Bow River, Alberta	..	3	1	6	4	6	9	7	3	3	2	1	6	3	3	2	2	1
Rest southern Alberta	..	4	2	3	1	3	9	7	1	4	5	1	1	1	1	1	–	–
British Columbia	..	–	1	2	–	4	5	5	3	–	4	4	4	3	2	2	1	–
Wyoming	..	–	2	1	1	3	1	2	2	1	1	1	1	2	1	–	2	1

	Dorsal rays								*Standard length (mm)*		
	N	*X̄*	*SD*	*CV*	*6*	*7*	*8*	*9*	*N*	*X̄*	*Range*
1892–1925 Cave & Basin	18	7.33	0.485	6.62	–	12	6	–	19	29.6	21.2–37.1
1920–40? Cave & Basin	79	7.63	0.511	6.70	1	27	51	–	84	34.2	18.8–46.8
1971–81 Cave & Basin	25	7.92	0.277	3.50	–	2	23	–	26	45.8	27.0–55.0
Bow River, Alberta	64	8.00	0.000	0.00	–	–	64	–	66	55.4	17.0–89.5
Rest southern Alberta	57	8.02	0.132	1.65	–	–	56	1	57	38.7	25.6–56.6
British Columbia	40	8.00	0.000	0.00	–	–	40	–	41	53.7	40.2–86.9
Wyoming	25	8.00	0.000	0.00	–	–	25	–	25	47.8	33.7–66.0

	*Abdominal vertebrae**									*Mean CV of L.L.,*
	N	*X̄*	*SD*	*CV*	*18*	*19*	*20*	*21*	*22*	*D., Ab. & Caud. Vert.*
1892–1925 Cave & Basin	19	19.53	0.772	3.95	1	9	7	2	–	5.30
1920–40? Cave & Basin	84	20.83	0.680	3.26	1	11	54	16	2	5.52
1971–81 Cave & Basin	25	19.80	0.577	2.91	1	4	19	1	–	3.96
Bow River, Alberta	66	19.68	0.660	3.35	1	24	37	3	1	3.37
Rest southern Alberta	56	19.64	0.645	3.28	–	24	29	2	1	3.65
British Columbia	41	19.95	0.740	3.71	–	11	22	7	1	3.49
Wyoming	22	20.09	0.526	2.62	1	2	16	3	–	3.56

	*Caudal vertebrae***										
	N	*X̄*	*SD*	*CV*	*17*	*18*	*19*	*20*	*21*	*22*	*23*
1892–1925 Cave & Basin	16++	20.00	0.816	4.08	–	–	4	9	2	1	–
1920–40? Cave & Basin	84	19.67	0.734	3.73	–	2	35	36	11	–	–
1971–81 Cave & Basin	25	19.76	0.723	3.66	–	–	9	14	1	1	–
Bow River, Alberta	63	19.78	0.728	3.68	–	1	22	30	10	–	–
Rest southern Alberta	57	19.67	0.636	3.23	–	1	21	31	4	–	–
British Columbia	41	20.97	0.908	4.33	–	–	2	10	17	11	1
Wyoming	22	18.82	0.733	3.89	1	5	13	3	–	–	–

Table 1. (Continued).

	Total vertebrae* **										
	N	*X̄*	*SD*	*CV*	*37*	*38*	*39*	*40*	*41*	*42*	*43*
1892–1925 Cave & Basin	19	39.58	0.692	1.75	–	1	7	10	1	–	–
1920–40? Cave & Basin	83	39.77	0.631	1.59	–	1	25	49	8	–	–
1971–81 Cave & Basin	24	39.58	0.881	2.23	1	–	10	10	3	–	–
Bow River, Alberta	63	39.46	0.737	1.87	–	4	31	23	5	–	–
Rest southern Alberta	55	39.31	0.663	1.69	–	4	32	17	2	–	–
British Columbia	41	40.93	0.787	1.92	–	–	1	10	22	7	1
Wyoming	22	38.91	0.610	1.57	–	5	14	3	–	–	–

* Including 4 centra for Weberian apparatus.
** Including urostyle.
++ No. not determinable accurately because of radiograph contrast.
SD = Standard deviation.
CV = Coefficient of variation.

reddish-orange nuptial color of the males have been recorded in Canada: Alouette River, B.C. (absent), Manitoba (present) by Bartnick (1972) and present in Lake Winnipeg and southern Ontario populations by Scott & Crossman (1973). Paetz & Nelson (1970) observed in the Cypress Hills and Red Deer tributaries of Alberta a reddish-orange tinge occurs at the base of the pectoral, pelvic, and anal fins and on the upper lip and lower cheek in breeding females, rather than in males.

The following breeding color records from NMNS records are given under provinces (arranged from west to east) and from north to south within a province, with catalog number (NMC), province, locality, date of collection, color notes, and water temperature. All records of 'color' were east of the Continental Divide.

British Columbia. – NMC81-0778: Buckinghorse River, Liard drainage, Mile 175 Alaska Highway (291.7 km), about 1 km west of bridge, 57°23′ N, 122°51′ W, 21 June 1978, some in spawning color with tubercles on head and pectoral fins.

Alberta. – NMC70-0114: Beaverlodge River, 2.4 km S and 6 km W of Huallen (Grande Prairie), 55°08′00″ N, 119°24′00″ W, 12 June 1969, fins were orange-red in color, 15.5° C.

Ontario. – NMC65-0133: Rideau River at Hogs Back, Ottawa, 45°22′39″ N, 75°42′00″ W, red behind lip and on pectoral and anal fin bases, 16.75° C.

ROM 14260: Ganaraska River in Durham County, 25 October 1943, 2 larger ones 70 mm long – all fins pink or reddish tinged most noticeable in case of pectoral and caudal and at base of pectoral, pelvic and anal fins; reddish on maxillaries.

Québec. – NMC63-0213: Nastapoka River, 3 km above mouth, about 82 km north of entrance to Richmond Gulf, 56°55′ N, 76°29′ W, dorsum orange colored.

NMC63-0221: Deer River, 3 km above mouth at Richmond Gulf, 56°17′ N, 76°08′ W, 19 July 1963, upper lip and lower fins red, 16.5° C.

NMC77-0703: Ascot River about 7 km NE of Waterville, Sherbrooke Co., 45°19′30″ N, 71°49′30″ W, 7 June 1977, in breeding color, 16° C.

NMC77-0700: Moe River about 3.5 air km NE of Compton, Compton Co., 45°15′30″ N, 71°47′30″ W, 7 June 1977, in breeding color, 16° C.

NMC77-0712: Ruisseau Chabot about 1 air km south of Saint-Isidore D'Auckland, 45°15′20″ N, 71°30′40″ W, 9 June 1977, in breeding color, 17° C.

NMC77-0713: Clifton River about 2 air km NW of Saint-Isidore D'Auckland, Compton Co., 45°16′30″ N, 71°32′30″ W, 9 June 1977, in breeding color, 17° C.

In summary males and sometimes females exhibit reddish or orangish red nuptial color east of the Continental Divide; this coloration is reportedly absent west of the Continental Divide.

Discussion

Chi-square tests showed that meristic values of early, mid and late samples were significantly different from one another. The mean values of the most diagnostic characters, lateral line scales and dorsal rays, increased over the study period. Mean percentage correctness of separation values for scales and dorsal rays also showed that early and late samples were the more different and that mid and late samples were the more similar.

It is clear that the three Banff Hotspring samples were not homogeneous, that there was a progressive shift in the value of the diagnostic meristic characters from early to mid to late samples and the late sample was indistinguishable from the Bow River sample. Because of unknown collecting methods for early and mid samples we do not think that size data can be used as evidence by itself, it is nevertheless worth noting that mean size displays a pattern concordant with values for scales and dorsal rays, that is the mean size increases from early through mid to late, 29.6, 34.2 and 45.8 mm SL, and Bow River sample averages somewhat larger, 55.4 mm SL.

The trend in mean values of lateral line scales and dorsal fin rays and the similarity of the late sample to Bow River stock leads us to postulate three hypotheses:

(1) Phenotypic hypothesis: the lateral line scale and dorsal fin ray counts in the samples were primarily determined by environmental conditions during ontogenetic development, and the environmental conditions changed over the sampling period. If so, the environmental factor must have shifted during the mid and late sampling periods and been most variable during the mid period.

(2) Genotypic hypothesis: the lateral line scale and dorsal fin ray counts were largely genetically determined and anthropogenic changes over the sampling period partially extirpated the indigenous hotspring dace and that Bow River stock (another subspecies) invaded the hotspring. Higher values of later samples could then have been due to:

(i) hybridization between hotspring and Bow River stocks,

(ii) or mixing of the individuals of the hotspring and Bow River stocks.

Phenotypic hypothesis

In regard to the first hypothesis, van Everdingen (1970 and unpublished tables 7A to 7D) demonstrated that there were seasonal variations in temperature and chemistry or outflows in Cave and Basin Hotsprings due to influx at the spring source of varying quantities of water having temperature and dissolved solid concentrations lower than the sulfurous thermal water. Over one annual cycle Cave Hotspring dropped from almost 34° C in April to almost 29° C in May, then warmed gradually over the rest of the year. The data for year to year variation are much sparser, however there seems to be no consistent trend. For example at Cave Spring in January 1917 the temperature was about the same (29.4° C) as on 17 March 1981 (29.3° C) with higher temperatures in 1967 and 1968 (17 March 1967 – 29.8° C, 10 October 1967 – 31.2° C, and 26 June 1968 – 31.0° C). The Basin Spring was highest in January 1916 – 34.4° C, the same on 27 May 1967 – 34.4° C, and lowest on 17 March 1981 – 33.5° C). Similar variation is seen in the chemistry of the two springs with no obvious trends present. The data suggest that seasonal variation equals between year variation, that there are greater differences between springs in one year than between years in a single spring, and that there has not been a general trend upward or downward in temperature or chemistry of the hotspring water. The available data do not support the phenotypic hypothesis. We are, however, unable to rule out some unknown sequence of environmental events that might have produced the changes in values and variance of meristics, except by recourse to Occam's razor.

We may note that it is not necessary to explain low fin ray and scale counts in *Rhinichthys* through phenotypic effects of high temperatures. In some cases, e.g. in *Rhinichthys osculus thermalis* (Hubbs & Kuhne 1937, Binns 1978), lower dorsal ray and lateral line counts *are* found in hotspring populations than in those living in nearby cooler streams. In other cases low scale counts (mean of 58.52 in Silvies River) are reported nearby to high counts (mean of 66.87 in Blitzen River) within the same river basin (Harney Basin, Oregon) for *Rhinichthys cataractae,* not in association with hotsprings (Bisson & Reimers 1977), and the authors argued that these two populations originated from different ancestral stocks. Hubbs et al. (1974, p. 108) observed that *R. osculus robustus* transplanted to a spring in Ruby Valley, Nevada, retained similar lateral line scale counts, which suggested genetic control. It is possible that the factors related to the small body size, typical of spring fishes, selected for low scale counts, in accordance with the pleomeristic rule (Lindsey 1975); however vertebral counts do not follow this pattern.

Genetic hypothesis

We will now consider the second general hypothesis that changes in morphological characters in Banff Hotspring dace were due to genetic changes, either as a result of hybridization with Bow River longnose dace, or as a result of admixture of the two stocks.

If the Banff longnose dace were undergoing hybridization with longnose dace from the Bow River, then the hotspring population would come to more closely resemble the Bow population. Comparison of mean lateral line counts of early, mid and late hotspring samples with Bow River samples show differences of 13 scales, 5 scales and 1 scale. Similarly, mean dorsal fin rays differed by 0.60, 0.37, and 0.08 rays. Clearly, the hotspring population became more like the Bow River population and the hybridization hypothesis survives this test, suggesting extinction through introgression.

If hybridization did take place between hotspring and Bow River populations, then one would expect an increase in variability due to mixing of the gene pools (Schueler & Rising 1976). Our coefficients of variation were greater for mid samples for lateral line scales and dorsal fin rays than for the early sample, supporting this contention; F-test showed the variance for lateral line scales was not quite significant ($F = 2.091$, df $1 = 81$, df $2 = 18$, $p = 0.05$–0.02). However, the variability for the late scale sample was less than the mid sample and only slightly higher than the early sample, and the late ray sample was less variable than the early or mid sample. Variability of the mid sample is in accordance with the hypothesis while that for the late sample is not.

Admixture hypothesis

It is however possible that the increase in mean values of the diagnostic meristic characters was due to an increasing proportion of Bow River stock amongst the hotspring population, without hybridization. A mixture of stocks in mid and late samples would also be expected to produce greater variance, unless one postulates that introgression was virtually complete in the late sample – a hypothesis supported by the indistinguishability of the late from the Bow River stock.

So it is necessary to find a test that distinguishes between hybridization and admixture. If the changes were solely due to admixture, then one should not find individuals in the mid and late samples with mixed diagnostic characters of hotspring and Bow River dace. Individuals with mixed characters would be expected on the other hand if there had been hybridization. Outside of the hotspring, Bow and other Alberta and B.C. specimens lacked 7 dorsal rays, which were fairly common in the hotspring. For specimens with 6 or 7 dorsal rays in the early sample, lateral line scales ranged from 51–59, in the mid sample 53–73, and in the late sample the only 7-rayed dace had 59 scales. Thus dace with 6 or 7 rays tended to have more lateral line scales in the mid and late samples than in the early sample. And most later specimens had more scales than any early 7-rayed specimen. A chi-square test shows this difference to be significant at $p<0.001$. We may conclude that individuals with mixed diagnostic characters were present in the mid and late samples, and therefore the admixture hypothesis may be rejected.

Were there nevertheless still some dace of pure or relatively pure stock in the mid and late hotspring samples, amongst the hybrids? Pure specimens would either have dorsal and scale counts both low or both high, while hybrids (and possible backcrosses) would tend to be intermediate. The calculated correlation between fin rays and scales in the mid sample was very low, 0.024, with $p>0.80$. The lack of correlation between the two characters suggests that most of the stock was of hybrid origin. Eight of 82 fish in the mid sample had 6 or 7 dorsal rays and fewer than 59 scales. These fish could be pure Banff longnose dace. However when two forms are hybridized some of the hybrids end up by chance having the diagnostic features of the parents, even though they are hybrids (and even though they may possess other unknown genes of the other parent). In the late sample only one specimen had 7 dorsal rays and 59 scales which might have been a Banff longnose dace.

In summary, the shift in mean values of the diagnostic characters over time and the mixture of diagnostic characters in most individuals in mid and late samples with possibly pure specimens in very low frequency suggests that the original population of Banff longnose dace underwent introgressive hybridization with Bow River longnose dace and that few if any of the original hotspring form remain.

Causes of decline and hybridization

Possible factors, many suggested by Lanteigne (1981), leading to a notable decrease in population of the Banff dace, include the following:

(1) The introduction of exotic species, and their subsequent multiplication to a population totalling in the thousands. Table 2 lists the 10 species and dates of introduction of exotic fishes which have been introduced either directly into the hotspring (9 species), or into park waters from whence they spread to the hotspring outflow (brook charr). Introduced species may have had a negative impact on native dace through:

(i) competition, e.g. for food, living space; (ii) predation on eggs, embryos, larvae or adults; (iii) aggression and territoriality; and (iv) introduction of diseases or parasites. The warm habitat may have permitted transmission of aquarium/tropical fish diseases or parasites to which the Banff longnose dace would have had little resistance.

(2) Pollution of the spring outflow. Human wastes from early heavy usage by bathers without adequate treatment facilities may have had negative effects on the dace environment. Dissolved oxygen levels of the springs are low, varying from 0 to 4.8 ppm (van Everdingen 1970). In the past the increased organic content may have increased the oxygen load and left critically low oxygen levels for eggs, young or adults. It is also possible that there were negative impacts from toxic substances that may have been used in the baths (chlorine?) or on the roads (de-icing salt?), or as pesticides. F.H. Schultz (personal communication to D.E.M.) reported the use of DDT in the 1950s when Schultz worked in the mountain parks.

(3) Alteration of outflow. Hotspring flow was probably altered during construction, repair, cleaning, and filling of the baths. Nelson (1983) cited a 1976 report mentioning a temporary diversion of hotspring outflow as being responsible for the disappearance of green swordtail and convict (= zebra) cichlids. Reductions or stoppages of flow could have had negative impacts by suddenly altering temperature, lowering water levels thus exposing them to avian predation, or stranding eggs, young or adults.

Table 2. Introduction of exotic fishes into Cave and Basin Hotsprings (or nearby).

Species	Date introduced	Date last observed
mosquitofish, *Gambusia affinis*	1924	September 1984
sailfin molly, *Poecilia latipinna*	autumn 1960	September 1984
guppy, *Poecilia reticulata*	before June 1968	before 1976
green swordtail, *Xiphophorus helleri*	before June 1968	before 1974
convict (= zebra) cichlid, *Cichlasoma nigrofasciatum*	before June 1968	before 1976
jewelfish, *Hemichromis bimaculatus*	before 1976	May 1981
angelfish, *Pterophyllum scalare*	1968	before May 1981
blue gourami (blue spot var.), *Trichogaster trichopterus*	? before 1968	died when flow shut off
Siamese fighting fish, *Betta splendens*	? before 1968	?
brook charr, *Salvelinus fontinalis*	?	September 1981

(4) A beaver dam presently backs up water into the marsh into which the hotsprings flow and has created a large pond (Fig. 2). It is not known when the beaver dam was constructed or whether there was a pond prior to the beaver dam. The dam was present and of about the same size in our earliest air photo taken 25 July 1950. It is possible that the construction of the beaver dam and the subsequent impoundment may have altered the primeval dace habitat. The dam was apparently built some time after 1918, and, see under History, was temporarily breeched and repaired in 1987.

It is difficult to evaluate which anthropogenic factor or factors caused the decline of the Banff longnose dace, although the introductions of exotics and interruptions in flow seem to us more important than pollution. Schoenherr (1981) discusses the roles of competition, predation, and aggression in the near extirpation and extinction of native fishes due to introduced species. However, despite the construction of the great bath-house and pool in 1914 and the introduction of mosquitofish in 1924 there were still pure Banff longnose dace present in 1925. Only about 32% of the 1920–40? sample were with any confidence assignable to Banff longnose dace, and at the very most 14% were so assignable by 1971–81 (assignment by possession of 48–58 lateral line scales, 6–7 dorsal rays, or both). The lowest lateral line scale count in the late sample (58 scales), taken in 1971, was at the borderline for the subspecies. It would therefore appear that most of the Banff longnose dace were extirpated/introgressed between 1925 and 1971. One of the ten introductions was made in 1924, and seven were probably made in the 1960s (Table 2) and probably resulted in the coup de grâce to the Banff longnose dace, even if cessations of flow had caused the initial decline.

Is restoration of *R. c. smithi* possible? It would be possible to conceive of a program to try and breed longnose dace more like *R. c. smithi*, by selecting the phenotypically least introgressed hybrids. Such a program would have to solve the problems of successfully raising and breeding dace and avoiding harmful affects of the introduced exotics. It must also face the problem of knowing only two characteristics, low dorsal ray and scale counts, of the Banff longnose dace. Specimens and some data are available to determine if there were other morphometric differences. Were there also biochemical and behavioral differences which adapted them to the hotspring environment? Solution of these problems by a recovery team just might be possible, given adequate resources. Observations by one of us (D.E.M.) while attending the winter Olympics in February 1988 to see his daughter ski, showed water levels in spring outflow less than 30 cm, suggesting the optimal time for a recovery program might be winter.

Taxonomic status

We have presented evidence suggesting that the native population of hotspring dace was reduced in numbers, that the spring was invaded by Bow River stock, that there was hybridization between the two stocks and that only the early sample, collected between 1892 and 1925, was uncontaminated by the Bow River population. Therefore, in determining the taxonomic validity of *Rhinichthys cataractae smithi*, we must use the early sample as representative, and exclude the mid and late samples.

In comparing early hotspring dace sample with Bow River sample we obtain nearly complete separation using lateral line scales. Nineteen of 19 early hotspring dace have 59 or fewer scales, while all but 3 of 62 the Bow River dace have 60 or more scales. This means that an average of 97.6% of these specimens are correctly assigned using this character (the average of correct assignment of each sample). The early hotspring sample is 89.2% distinguishable from the rest of southern Alberta sample, and 93.6% separable from all non-hotspring Alberta longnose dace combined. From the B.C. sample the mean separation is 98.8%, and from the combined non-hotspring Alberta and B.C. sample the separation is 90.1%. The holotype of *R. c. smithi* has 52 lateral line scales.

In regard to dorsal fin rays 66.7% of early Banff hotspring dace had 7 dorsal rays, the rest 8. One hundred percent of the Bow, the rest of Alberta and the B.C. specimens had 8 dorsal rays, except for one southern Alberta specimen which had 9 rays. The holotype has 7 dorsal rays. The dorsal rays, by themselves, or in combination with lateral line scales, were useful in distinguishing some specimens of *R. c. smithi*, but not highly diagnostic. Size and some other morphometric characters showed differences between early hotspring and other samples but we do not analyze them here.

Bailey et al. (1954) suggested that before subspecies be recognized that a level of difference of at least 93% was desirable; other authors (Mayr 1969) suggested lower values. The percentage separation level of early Banff longnose dace is at or above that expected for subspecies recognition. We also note that the differences between mean lateral line scale counts of hotspring, and neighbouring Alberta and B.C. samples, are greater than those found between our stream samples and those in Bisson & Reimers (1977). The holotype has the diagnostic characters of the hotspring subspecies. We therefore conclude that *Rhinichthys cataractae smithi* Nichols, is a valid subspecies.

Relationships of Rhinichthys cataractae smithi

The *Rhinichthys cataractae* complex is in a taxonomically confused state. It is not our intention to undertake a revision, but we will try to establish to which of the putative subspecies of *R. cataractae* the Banff longnose dace is most closely related and what is its scientific name.

The nominate subspecies, *R. c. cataractae* (Valenciennes *in* Cuvier & Valenciennes, 1842) was described from Niagara Falls. It occurs, according to Hubbs & Lagler (1964), in the Great Lakes drainage basin, ranging around all the shores of the Great Lakes, and also into cool, swift streams in the northern two-thirds of Wisconsin, in Michigan northward from the Muskegon and Au Sable rivers, and in the St. Joseph system of northwestern Indiana; down the St. Lawrence to the region of Montreal. They also stated that the 'Relationship to other subspecies in characters and distribution [was] not yet precisely determined ...'.

Simon (1951) in *Wyoming fishes* described *Rhinichthys cataractae ocella* Garman, 1881 as present in all waters of the upper Missouri, excepting the upper North Platte drainage, as well as occupying the Columbia River and Bonneville basins (one of the few forms of fishes, he stated, that occurs on both sides of the Continental Divide). However in a revised edition of *Wyoming fishes* Baxter & Simon (1970) no longer recognized the subspecies *R. c. ocella*, nor *R. c. dulcis*. The type-locality of *R. c. ocella* is northwestern Wyoming and Montana.

Girard (1857) originally described *Rhinichthys cataractae dulcis* (Girard, 1857) (in the genus *Argyreus* as a full species) from the Sweetwater River in Wyoming (not in Nebraska as listed by Jordan et al. 1930, as pointed out by R.R. Miller, *in litt.* 21 Nov. 1985). Girard also stated that smaller specimens of the same species were obtained in Cottonwood Creek, Utah. Jordan et al. (1930) gave the range of *R. dulcis* (with *R. ocella* treated as a synonym) as: upper Missouri, Platte, Arkansas, Rio Grande and Colorado rivers; Columbia River; Utah Basin; and coastwise streams of Washington and Oregon.

McPhail & Lindsey (1970) noted that breeding males of *Rhinichthys cataractae* east of the Continental Divide [in Canada] develop reddish coloration bordering the upper lip and at the base of each of the lower fins and that west of the Divide the color is faint or absent. Bartnik (1972) distinguished two subspecies of longnose dace in western Canada on the basis of spawning color and behavior. Following earlier authors, Bartnick (1972) described the range of *R. c. cataractae* as east of the Continental Divide and *R. c. dulcis* as on both sides of the Continental Divide. He ascribed studied specimens from the Alouette River in southwestern B.C. to *R. c. dulcis* and from Manitoba to *R. c. cataractae*. On this basis he reported that *R. c. cataractae* males display bright crimson patches on the ventral surface (angles of lips, cheeks, bases of paired fins and anal fin), females show light pink color in the same areas, and that they breed during daylight, while in *R. c. dulcis* males this red color is faint or lacking, in females there is no breeding color whatsoever, and they court and breed at night. Our own data support the existence of eastern and western color morphs and extend their known distribution.

We summarize our tentative views on classification of western *Rhinichthys cataractae* populations in the key below.

The westslope longnose dace, with uncolored spawning males, appears restricted to the Pacific drainage west of the Rockies, except possibly for the upper Peace River in the Arctic Basin. It is unknown whether the different times of courtship between westslope and eastslope forms are sufficient to constitute a reproductive isolation. We have not found an available name for this form and if it is found to be distinct it will have to be described as a new species or subspecies.

We note that Girard (1857) described *Argyreus* (= *Rhinichthys*) *dulcis* as having the anterior margin of the dorsal fin nearer to the extremity of the snout than to the insertion of the caudal fin. Ten specimens of 10 examined from the type-locality of *A. dulcis* had the dorsal origin behind the midpoint of the body and a small sample taken east and west of the Continental Divide showed the dorsal origin in *Rhinichthys cataractae* to be typically behind the midpoint of the body or occasionally at or near the midpoint.

The eastslope longnose dace is found on the eastern slope of the Rockies. Presumably it is this form which ranges north to the lower Mackenzie River, Peel and lower Peace systems and this is supported by the spawning color of males taken in the Liard drainage of British Columbia (Results). It was unknown how far east and south this form ranges or whether it is consubspecific with the nominate subspecies, *Rhinichthys cataractae cataractae*. If an eastslope subspecies is distinguishable from the nominate subspecies then the oldest available name for the eastslope form would be *Rhinichthys cataractae dulcis* (Girard, 1857). Hubbs (1926) had earlier made comparisons of series of longnose dace specimens from various parts of its extensive range and found that characters supposed to set off an eastslope subspecies, *R. c. dulcis,* do not hold. It is possible that a future multivariate study or discovery of new characters might show support for recognition as *R. c. dulcis* of those longnose dace that are hypothesized to have survived glaciation in the upper Missouri drainage (Bailey & Allum 1962, Crossman & McAllister 1986). But as neither previous published studies nor our own observations support recognition of this subspecies we presently regard *Argyreus dulcis* Girard, 1857 as consubspecific with and a subjective junior synonym of *Gobio cataractae* Valenciennes *in* Cuvier and Valenciennes, 1842. We note that the high scale count (70) and large size (5 pouces = 127 mm) in the original description make clear that Valenciennes had the many-scaled eastern form of *Rhinichthys cataractae* and not *R. c. smithi* or *Rhinichthys atratulus* (Hermann, 1804). Thus the correct name for longnose dace in all of Canada except for the Banff hotsprings and the Pacific Basin and the upper Peace River is *Rhinichthys cataractae cataractae* (Valenciennes, *in* Cuvier & Valenciennes, 1842).

For the moment, the two putative subspecies parapatric to or closest to *Rhinichthys cataractae smithi* are the unnamed westslope longnose dace and the eastern longnose dace, *R. c. cataractae.* The male nuptial color pattern of the Banff longnose dace is unknown although male hybrids collected by Nelson (1983) 16–17 May 1981 were orange-colored in part of the snout and mouth area, indicative of the eastern longnose dace. The mean lateral line scale count of the Banff longnose dace, 54.6 tends to be closer to eastslope samples of 65.8 (Bow River), 63.7 (rest of Alberta) than to westslope means of 66.8 (B.C.), or the two closest U.S. Columbia River Basin samples (Bisson & Reimers 1977) of 67.8 (Palouse River, Idaho) and 64.9 (Grande Ronde River, southeast Washington) [scale counts further downstream in the Columbia are lower]. The modal vertebral counts (Table 1), further suggest similarity to eastern rather than westslope longnose dace.

The meagre available meristic data support relationships of Banff longnose dace to the eastern rather than the westslope longnose dace. As the Cave and Basin Hotsprings are on the east slope of the Rockies, this provisional conclusion is in conformity with topographic expectations and does not require stream piracy from the Pacific Basin.

But despite topographic expectations it is possible that the Banff longnose dace is related to allopatric populations of the longnose dace complex of the Pacific Northwest. Most of the Washington, Oregon and Idaho populations of longnose dace studied by Bisson & Reimers (1977) were similar to westslope longnose dace in B.C. and were distinguished from Banff longnose dace by high lateral line scale counts (population means from 60.41–67.83) or high dorsal counts (7.90–9.00), or both. A comparison of those populations with lower counts follows using Bisson & Reimers (1977) data. The John Day River, Oregon sample had a mean lateral line count of 59.40 and a dorsal ray mean of 7.9, thus differing by more than 5 scales and half a ray. The Silvies River, Oregon sample had a mean of 58.52 scales and 8.0 dorsal rays, thus differing by 4.8 scales and 0.6 rays. The Chehalis River, Washington sample (= Nooky longnose dace) with 57.61 scales and 8.0 rays differed from Banff longnose dace by almost 4 scales and 0.6 dorsal rays; but another sample from the Chehalis River reported by McPhail (1967) had a mean of only 53.8 scales, very close to that for Banff dace. McPhail (personal communication) reported that the nuptial color of male Nooky longnose dace is pale as is typical of westslope longnose dace. The Umpqua, Oregon population (= *Rhinichthys evermanni* Snyder, 1908) with 58.96 scales and 9.0 rays differed by 5.3 scales and 1.6 dorsal rays. To summarize, all Pacific Northwest samples (except one, the Nooky longnose dace) differed by means of at least 4.98 scales and at least 0.5 rays (7 was a rare count in all populations), and all attained a maximum SL at least one and a half times greater. None were reported to inhabit hot springs. All relatively low scale and dorsal ray count populations were at least 1000 air km (further by river routes) distant and all are in the Pacific as opposed to the Hudson Bay Basin. Intervening populations in B.C. and Idaho have higher counts. We conclude that the moderately low scale and fin ray counts found in some populations of the Pacific Northwest had an independent origin and are taxonomically distinct from *R. cataractae smithi*.

Origin of the Banff longnose dace

With few exceptions such as Beringian Yukon, Canada has traditionally been considered to have almost completely been covered by the Wisconsin glacial sheets. According to this view *Rhinichthys cataractae smithi* would have originated postglacially. According to the 1969 Geological Survey of Canada Map 1257A, Retreat of Wisconsin and recent ice in North America, the Banff area would have become deglaciated at about 9500 years before present. Origin of a subspecies in such a time frame is within accepted rates of evolution.

However, according to Reeves (1973) and Crossman & McAllister (1986) there may have been an icefree refugium on the east slope of the Rockies in the environs of Banff and Jasper national parks. The biological evidence consists of a number of endemic taxa of molluscs, crustaceans, fishes, and plants which are restricted to this region. However one could propose a smaller special refugium for the hotspring dace based on local melting of the icesheet by hotsprings on Sulphur Mountain. In support of such a local refugium would be the occurrence of the snail *Physa johnsoni*, endemic to hot and cold springs and their effluent streams in Banff National Park. But other eastslope endemics more widespread than this would appear to call for a larger refugium. Crossman & McAllister (1986) discuss geological evidence pro and con such a refugium.

Minimally it would appear that *R. c. smithi* would have evolved postglacially from populations which survived glaciation in southern Alberta or Montana in the upper Missouri drainage basin, and be younger than 9500 BP. Or maximally this subspecies may have survived the Wisconsin glaciation in situ in a local or larger eastslope Alberta refugium and be much older.

Preliminary Key to *Rhinichthys cataractae* complex in British Columbia, Alberta, Washington and Oregon

1 (6) Breeding males lack or possess only faint red color at angles of lips, cheeks and bases of lower fins and completely lacking in females. Courtship and spawning nocturnal. Total vertebrae modally 40–41. Pacific Basin and perhaps upper Peace drainage.
2 (5) Dorsal rays 8, rarely 7.
3 (4) Lateral line scales usually 58–75, scales around caudal peduncle 27–34. Columbia and Fraser (? upper Peace) river drainages

westslope longnose dace – *Rhinichthys cataractae* undescribed subspecies.

4 (3) Lateral line scales 50–61, scales around caudal peduncle 21–28. Chehalis River and Puget Sound drainages, Washington north to Nooksack River, British Columbia .... Nooky longnose dace – *Rhinichthys cataractae* subsp.[1]

5 (2) Dorsal rays 9. Umpqua River, Oregon ..... Umpqua longnose dace. *Rhinichthys evermanni* Snyder, 1908.[2]

6 (1) Breeding males with orange or bright crimson color at angle of lips, cheeks and bases of lower fins; females with light pink in same areas.[3] Courtship and spawning diurnal. Total vertebrae 39–40. Occur in Atlantic, Arctic and probably Gulf of Mexico drainage basins.

7 (8) Lateral line scales usually 50–58 (48–59). Dorsal rays 7–8. Maximum SL less than 40 mm SL. Occurred in Cave and Basin Hotsprings, Banff National Park, Alberta .... Banff longnose dace – *Rhinichthys cataractae smithi* Nichols, 1916.

8 (7) Lateral line scales usually 59–73 (56–75). Dorsal rays 8(9). Maximum SL exceeds 90 mm. Ranges east from the Continental Divide to at least the Great Lakes-Saint Lawrence River Basin and north to the lower Mackenzie, Peel and Liard drainages. .... eastern longnose dace – *Rhinichthys cataractae cataractae* (Valenciennes *in* Cuvier & Valenciennes, 1842).

Acknowledgements

R.R. Campbell, Fisheries and Oceans Canada, Ottawa, made available a grant of $2000 to support this study, encouraged and facilitated its completion. Study material was kindly lent to us by E.J. Crossman, Royal Ontario Museum; W. Eschmeyer, California Academy of Sciences; R.R. Miller, Museum of Zoology, University of Michigan; Joseph S. Nelson, University of Alberta; and A.P. Russell, University of Calgary. Clarence Smith, American Museum of Natural History, permitted one of us (C.B.R.) to examine the holotype and paratypes of *Rhinichthys cataractae smithi* in his care. Jacqueline Lanteigne, assisted by Fern McAllister, collected specimens and made field observations; she also read the manuscript. Peter A. Bisson, Weyerhauser Co., Tacoma, Washington, Brian W. Coad, National Museum of Natural Sciences, E. Kott, Wilfrid Laurier University, C.C. Lindsey, University of British Columbia, and J.S. Nelson, University of Alberta critized the manuscript. R.R. Miller provided useful supplementary information on dace. J.D. McPhail, University of British Columbia, permitted us to quote him regarding the lack of nuptial color in male Nooky dace, and Erling Holm of Royal Ontario Museum dug out some 45 year old data of nuptial color from Ontario longnose dace. We appreciate information supplied by Dave Mayhood on the 1987 draining of the Cave and Basin Pond; and his comments on the manuscript. To all these persons the authors are very grateful.

Appendix

The following information is quoted from a biography by W. McCardell written late in life. He, his brother, and Frank McCabe, railway workers, claimed that they discovered the hotsprings in November 1883. We have not edited the text.

'Soon we were climbing up the strange formation of which the bank was composed. Once on top the first bench we rushed around to try and locate the place were the warm water came from. This did not take long. Out of a beautiful circular basin, jamed full of logs all the way in size from four to eight inches in diameter, some ten inches, seeped this warm water. Our enthusiastic interest at this discovery was beyond the description of words. Our joyousness and invigorating thrill we experienced in thus locating this strange phenomena, and hidden secret of the wilderness, knew no bounds.

During our stay of perhaps twenty minutes, looking at the circular basin, it suddenly dawned upon us, that there might be other places of just as much interest as this pool which threw out this great volume of warm sulphur water. For such it proved to be.

So we moved on up to another bench and to our utter astonishment we came upon a great cave 35 or 40 feet in depth. This was the greatest climax to a major discovery, that we had ever seen. Frank McCabe, Tom McCardell and myself just stood in silence looking at this mysterious grotto where warm clear water

[1] For further data on this undescribed species/subspecies see McPhail (1967, 1980), Bisson & Reimers (1977), McAllister et al. (1985).

[2] Treated as a full species by Lee et al. (1980).

[3] Nuptial color unknown for *R. c. smithi.*

bubbled from its depths. At last Tom McCardell spoke, saying 'what are we going to do now?'

'Why continue our search? Who knows,' said Tom in a lighter vein, 'but that this is the fountain of youth.'

'We don't know what secret of nature's treasure may be revealed in this spot. We will just hunt around a few hundred feet from where we first landed, then we will return to this same cavern and decide what action we will take to search this mysterious basin. That is so far as our skill will permit,' Tom added as an afterthought.

Thus we started off in different directions. Tom McCardell went east, while Frank McCabe and myself went west. We had hardly started off in our different directions of search, when we shouted to Tom,

'Come and see the little fish swimming in this stream!' This little stream flowed out of a small cave not far from the big cave opening. This little stream of warm water followed along the bench for about one hundred feet and tumbled over the bench face into the circular basin that we had just recently left to come up to the cave mouth. We drove these little minnows down towards the falls that went into the basin, but when they came near the falls, they persistently turned back up stream in spite of all our efforts to get them to go over the falls.

It was evident that these peculiar fish belonged to this little stream of warm water. We proved this by placing some in a vessel until the water got cold. When the temperature of the water dropped these little fish would die. On other occasions afterwards when we took water out of the stream to cool for drinking purposes, and inadvertantly scooped some of the small minnows with this water, we would find these fish dead in the morning. We decided to eat our lunch and do further exploring some other day.

The mystery of the cave however still intrigued us, and we decided that one of the party would go over to the wigwam and get a rope to lower some one down to the bottom of the cave and see what it was like below.

We ate our lunch but did not make tea. We just drank the sulphur water out of the little stream which sheltered in its warm waters the little fish which skittered about. How long they had been in this water, and what processes of evolution took place in order for them to adapt themselves to this warm environment, no one seems to have any idea.

I visited my old haunts recently, and I saw the same species of fish in this water as I had seen over 50 years ago, when we made our voyage of discovery.'

References cited

Anonymous. 1987. Fishy incident at the Cave. Canadian Parks & Wilderness Society Newsletter 4(1): 14.

Bailey, R.M. & M.D. Allum. 1962. Fishes of South Dakota. Museum of Zoology, University of Michigan, Miscellaneous Publications (119): 1–131.

Bailey, R.M., H.E. Winn & C.L. Smith. 1954. Fishes from the Escambia River, Alabama and Florida, with ecological and taxonomic notes. Proc. Acad. Nat. Sci. USA 106: 109–164.

Bartnik, V.G. 1972. Comparison of the breeding habits of two subspecies of longnose dace, *Rhinichthys cataractae*. Can. J. Zool. 50: 83–86.

Baxter, G.T. & J.R. Simon. 1970. Wyoming fishes. Wyoming Game and Fish Department Bulletin, Cheyenne (4): 1–168.

Binns, N.A. 1978. Habitat structure of Kendall Warm Springs, with reference to the endangered Kendall Warm Springs dace, *Rhinichthys osculus thermalis*. Wyoming Fish and Game Department Fisheries Technical Bulletin (4): 1–45.

Bisson, P.A. & D.E. Reimers. 1977. Geographic variation among Pacific Northwest populations of longnose dace, *Rhinichthys cataractae*. Copeia 1977: 518–522.

Butcher, G.A. 1980. Multivariate analysis of hybridizing Cyprinidae from the Kananaskis reservoirs, Alberta. Can. J. Zool. 58: 1664–1673.

Campbell, R.R. 1985. Rare and endangered fishes and marine mammals of Canada status reports. II. Can. Field-Naturalist 99: 404–408.

Crossman, E.J. 1984. Introduction of exotic fishes into Canada. pp. 78–101. *In:* W.R. Courtenay & J.R. Stauffer (ed.) Distribution, Biology and Management of Exotic Fishes, The John Hopkins University Press, Baltimore.

Crossman, E.J. & D.E. McAllister. 1986. Chapter 3. Zoogeography of freshwater fishes of the Hudson Bay drainage, Ungava Bay and Arctic Archipelago. pp. 53–104. *In:* E. Wiley & C.H. Hocutt (ed.) Zoogeography of the Freshwater Fishes of North America, John Wiley, New York.

Cuvier, G.L.C.F.D. & A. Valenciennes. 1842. Histoire naturelle des poissons. P. Bertrand, Paris. 472 pp.

Dymond, J.R. 1947. A list of the freshwater fishes of Canada east of the Rocky Mountains with keys. Royal Ontario Museum of Zoology, Toronto. 36 pp.

Eigenmann, C.H. 1894. Results of explorations in western Canada and the northwestern United States. Bulletin of the United States Fish Commission for 1894 14: 101–132.

Girard, C. 1857. Researches on the cyprinoid fishes inhabiting the fresh waters of the United States of America, west of the Mississippi Valley, from specimens in the museum of the Smithsonian Institution. Proc. Acad. Nat. Sci. USA 8: 165–213.

Hubbs, C.L. 1926. A check-list of fishes of the Great Lakes and tributary waters, with nomenclatorial notes and analytical keys. University of Michigan Museum of Zoology, Miscellaneous Publications (15): 1–77.

Hubbs, C.L. & E.R. Kuhne. 1937. A new fish of the genus *Apocope* from a Wyoming warm spring. Occasional Papers of the Museum of Zoology, University of Michigan (343): 1–21.

Hubbs, C.L. & K.F. Lagler. 1964. Fishes of the Great Lakes region. University of Michigan Press, Ann Arbor. 213 pp.

Hubbs, C.L., R.R. Miller & L.C. Hubbs. 1974. Hydrographic history and relict fishes of the North-Central Great Basin. Memoirs of the California Academy of Sciences 7: 1–259.

Jordan, D.S., B.W. Evermann & H.W. Clark. 1930. Check list of the fishes and fishlike vertebrates of North and Middle

America north of the northern boundary of Venezuela and Columbia. Appendix X, Report of the United States Commissioner of Fisheries for the fiscal year 1928. 670 pp.

Lanteigne, J. 1981. Status of the Banff longnose dace, *Rhinichthys cataractae smithi*. Report to the Committee on the Status of Endangered Wildlife in Canada. 20 pp.

Lee, D.S., C.R. Gilbert, C.H. Hocutt, R.E. Jenkins, D.E. McAllister & J.R. Stauffer, Jr. 1980. Atlas of North American freshwater fishes. North Carolina State Museum of Natural History, Raleigh. 854 pp.

Lindsey, C.C. 1975. Pleomerism, the widespread tendency among related fish species for vertebral number to be correlated with maximum body length. J. Fish. Res. Board Can. 32: 2453–2469.

Mayr, E. 1969. Principles of systematic zoology. McGraw-Hill Book Company, Toronto. 428 pp.

McAllister, D.E. 1969. Introduction of tropical fishes into a hotspring near Banff, Alberta. Can. Field-Naturalist 83: 31–35.

McAllister, D.E. 1970. Rare or endangered Canadian fishes. Can. Field-Naturalist 84: 5–8.

McAllister, D.E. & C.G. Gruchy. 1977. Status and habitat of Canadian fishes in 1976. pp. 151–157. *In:* T. Mosquin & C. Suchal (ed.). Canada's threatened species and habitats. Proceedings of the Symposium on Canada's Threatened Species and Habitats co-sponsored by the Canadian Nature Federation and the World Wildlife Fund (Canada). 185 pp.

McAllister, D.E., B.J. Parker & P.E. McKee. 1985. Rare, endangered and extinct fishes of Canada. National Museum of Natural Sciences, Ottawa, Syllogeus (54): 1–192.

McPhail, J.D. 1967. Distribution of freshwater fishes in western Washington. Northwest Science 41: 1–11.

McPhail, J.D. 1980. Threatened and endangered species and habitats in British Columbia and the Yukon. pp. 139–145. Province of British Columbia Ministry of the Environment.

McPhail, J.D. & C.C. Lindsey. 1970. Freshwater fishes of northwestern Canada and Alaska. Fisheries Research Board of Canada Bulletin (173): 1–381.

Miller, R.R. 1984. *Rhinichthys deaconi,* a new species of dace (Pisces: Cyprinidae) from southern Nevada. Occasional Papers of the Museum of Zoology, University of Michigan (707): 1–21.

Morgan, R.P., II, W.L. Goodfellow, C.H. Hocutt & J.R. Stauffer, Jr. 1984. Karyotype of *Nocomis micropogon, Rhinichthys cataractae*, and their supposed hybrid, *'Rhinichthys bowersi'* (Pisces: Cyprinidae). Copeia 1984: 990–992.

Nelson, J.S. 1983. The tropical fish fauna in Cave and Basin Hotsprings drainage, Banff National Park, Alberta. Can. Field-Naturalist 97: 255–261.

Nelson, J.S. & M.J. Paetz. 1982. Alberta fish species. Alberta Naturalist 12: 52–53.

Nichols, J.T. 1916. On a new race of minnow from the Rocky Mountains Park. Bulletin of the American Museum of Natural History 35: 69.

Paetz, M.J. & J.S. Nelson. 1970. The fishes of Alberta. The Queen's Printer, Edmonton. 282 pp.

Reeves, B.O.K. 1973. The nature and age of the contact between the Laurentide and Cordilleran ice sheets in the western interior of North America. Arctic and Alpine Research 5: 1–16.

Schoenherr, A.A. 1981. The role of competition in the replacement of native fishes by introduced species. pp. 173–203. *In:* R.J. Naiman & D.L. Soltz (ed.) Fishes in North American Deserts, John Wiley & Sons, Toronto. 552 pp.

Schueler, F.W. & J.D. Rising. 1976. Phenetic evidence of natural hybridization. Systematic Zoology 25: 283–289.

Scott, B.W. & E.J. Crossman. 1973. Freshwater fishes of Canada. Fisheries Research Board of Canada Bulletin (184): 1–966.

Simon, J.R. 1951. Wyoming fishes. Revised. Wyoming Game and Fish Department Bulletin, Cheyenne (4): 1–124.

Slastenenko, E.P. 1958. The freshwater fishes of Canada. Kiev Printers, Toronto. 385 pp.

Smyth, D. 1985. The Cave and Basin Hot Springs. Beaver, Winter 1984/85, Outfit 315: 3. pp. 46–51.

Stauffer, J.R., Jr., C.H. Hocutt & R.F. Denoncourt. 1979. Status and distribution of the hybrid *Nocomis micropogon* X *Rhinichthys cataractae*, with a discussion of hybridization as a viable mode of vertebrate speciation. Amer. Midl. Nat. 101: 355–365.

van Everdingen, R.O. 1970. Seasonal variations, Sulphur Mountain Hot Springs, Banff, Alberta. Canada Department of Energy, Mines and Resources, Inland Waters Branch, Ottawa, Technical Bulletin (33): 1–11.

Vadim Vladykov at the 1974 ASIH meeting in Ottawa, as usual skeptical about orthodox ideas.

Environmental Biology of Fishes Vol. 23, No. 1–2, pp. 115–125, 1988

Aphanius vladykovi, a new species of tooth-carp from the Zagros Mountains of Iran (Osteichthyes: Cyprinodontidae)

Brian W. Coad
Ichthyology Section, National Museum of Natural Sciences, P.O. Box 3443, Station D, Ottawa, Ontario, Canada K1P 6P4

Received 30.11.1987 Accepted 23.12.1987

Key words: Orogenic events, Isolation, Pupfish, Taxonomy

Synopsis

A new species of tooth-carp, *Aphanius vladykovi*, is described from the Zagros Mountains of Iran. The new species is uniquely characterised by a high count of lateral line scales and by different pigmentation patterns from its closest, putative relative *A. sophiae*. It also possesses several meristic and morphometric differences with topotypic *A. sophiae*. It became separated from a common ancestor with *A. sophiae* by orogenic events which led to the formation of the Zagros Mountains and the isolation of the upper reaches of the Karun River basin by the development of *tangs*, narrow passes through the mountains with high water flow and steep gradient which prevent gene flow between lowland and mountain populations of these small fishes.

Introduction

The Cyprinodontidae are represented in Iran by four species. These are *Aphanius ginaonis* (Holly, 1929) restricted to a hot spring near Bandar Abbas at the Straits of Hormuz, *A. mento* (Heckel, 1843) which is found in the Tigris River basin in Iran, *A. dispar* (Rüppell, 1828) found in all the coastal drainages of the Persian Gulf and the Sea of Oman and in inland basins of Baluchestan, and *A. sophiae* (Heckel, 1849) found in the Tigris River basin, rivers draining to the northern Persian Gulf, and certain internal basins, namely the Kor River basin, the Lake Maharlu basin, possibly the Esfahan basin, and the Namak Lake basin (Coad 1987). This paper describes a fifth, new species from the Zagros Mountains in the upper reaches of the Karun River drainage which empties at the head of the Persian Gulf.

Materials and methods

Counts and measurements follow Hubbs & Lagler (1958). Gill raker counts are the total number of rakers on the first gill arch and fin ray counts are the total number of rays in each fin. Vertebrae counts include the hypural plate as one vertebra. Measurements which include the anterior end of the fish are to the snout tip, not the protruding lower jaw. Dorsal and anal fin lengths are from the fin origin to the posteriormost tip of the fins. Scale counts, except those in lateral series, are given as ranges only since scales are strongly embedded in places or overlain by thick skin in the new species and accurate counts are not possible without staining. Counts for these characters agree generally with those made on alizarin-stained material.

Abbreviations: SL = standard length; HL = head length; NMC = National Museum of Natural Sciences, National Museums of Canada, Ottawa;

NMW = Naturhistorisches Museum, Wien.

The specimens on which the description is based were as follows: Holotype: NMC 79-0247, male, 36.6 mm SL, Iran, Shahrestan-e Bakhtiari va Chahar Mahall, large pool, 3 km west of Boldaji, 31° 57′ N, 51° 01′ E, 9 June 1977, B.W. Coad and S. Mansoorabadi. Paratypes: NMC 79-0247A, 35 males 21.6–36.4 mm SL and 16 females 23.5–40.2 mm SL, same locality as holotype. NMC 79-0248, male, 30.3 mm SL, Iran, Shahrestan-e Bakhtiari va Chahar Mahall, stream, 3 km east of Boldaji, 31° 55′ N, 51° 05′ E, 9 June 1977, B.W. Coad and S. Mansoorabadi.

Comparative material: *Aphanius sophiae,* NMC 79-0025, female 31.6 mm SL, Iran, Fars, Kor River at Marv Dasht, 29° 51′ N, 52° 46′ 30″ E; NMC 79-0059, 19 males 18.1–39.3 mm SL and 11 females 24.5–31.9 mm SL, Iran, Fars, Pulvar River, 8 km south of Sivand, 30° 01′ 30″ N, 52° 57′ E, 18 April 1976, B.W. Coad and S. Coad; NMC 79-0061, 3 females 23.2–32.1 mm SL, Iran, Fars, stream tributary to Pulvar River, 14 km south of Sa'adatabad, 30° 04′ N, 53° 01′ E, 18 April 1976, B.W. Coad and S. Coad; NMC 79-0062, 2 males 24.0–26.6 mm SL and 4 females 26.5–42.0 mm SL, Iran, Fars, spring, 17 km south of Sa'adatabad, 30° 05′ N, 53° 00′ E, 18 April 1976, B.W. Coad and S. Coad; NMC 79-0067, male 25.7 mm SL, Iran, Fars, qanat at Zarqan, ~29° 46′ N, ~52° 43′ E, 27 April 1976, B.W. Coad and S. Coad; NMC 79-0292, 4 males 26.5–31.1 mm SL and 2 females 29.7–34.4 mm SL, Iran, Fars, Lapu'i spring near Zarqan, 29° 48′ N, 52° 39′ E, 30 June 1974, H. Assadi; NMC 79-0342, female 39.4 mm SL, Iran, Fars, Kor River at Band-e Amir, 29° 49′ N, 52° 51′ E, 22 November 1977, B.W. Coad; NMC 79-0498, 4 males 19.7–25.2 mm SL and 2 females 17.4–20.0 mm SL, Iran, Fars, spring in Kor River basin, 30° 05′ N, 52° 27′ E, 4 October 1978, B.W. Coad.

All the above material of *A. sophiae* is from the drainage basin of the Kor River, the type locality of this species. All specimens were counted and measured for Tables 1 to 3 except the smallest female in NMC 79-0498. The following material comprises part of an extensive type series of specimens and nominal species referred to *A. sophiae.* This material was in some instances decoloured and all the specimens had shrunk noticeably since they were catalogued in 1844 compared to relatively fresh material at hand from the type locality. Accordingly, although they were examined to confirm descriptions of this species they were not measured for inclusion in Tables 1 to 3.

Lebias sophiae (syntypes of *Aphanius sophiae*): NMW 60327 (8 specimens), 25.2–31.1 mm SL, Iran, Fars, 'Salzquellen v. Persepolis', Th. Kotschy; NMW 22616/623, (8), 26.1–29.0 mm SL, as above; NMW 75067 (7), 24.7–27.4 mm SL, Iran, Fars, 'Persepolis', Th. Kotschy; *Lebias crystallodon* (holotype and synonym of *A. sophiae*), NMW 15175, 40.1 mm SL, Iran, Fars, 'Nemek Deira [*sic*] bei Persepolis', Th. Kotschy; *Lebias punctatus* (syntypes and synonym of *A. sophiae*), NMW 15070 (5), 29.9–35.6 mm SL, Iran, Fars, 'Nemek Deira [*sic*] bei Persepolis', Th. Kotschy; NMW 59609 (6), 21.4–34.7 mm SL, as above; NMW 59837 (5), 26.6–35.7 mm SL, as above; NMW 15156 (14), 19.2–32.1 mm SL, Iran, Fars, 'Salzquellen von Persepolis', Th. Kotschy.

Taxonomy

Diagnosis

The new species is clearly distinguished from all other members of the genus *Aphanius* sensu stricto in Iran and the Middle East by the higher number of lateral line scales (36–47). Other members of the genus *Aphanius,* and the related *Kosswigichthys,* have 32, usually 30 scales or less in lateral series (Jenkins 1910, Berg 1949, Villwock 1964, Parenti 1981, Villwock et al. 1983, Tortonese 1986). Male colour on the dorsal fin is much darker and the anal fin is light compared to *A. sophiae,* the species which most closely resembles the new species. Females lack the typical large spot at the base of the caudal fin found in *A. sophiae.* There are also further differences in meristic and morphometric characters which are demonstrated statistically although ranges overlap.

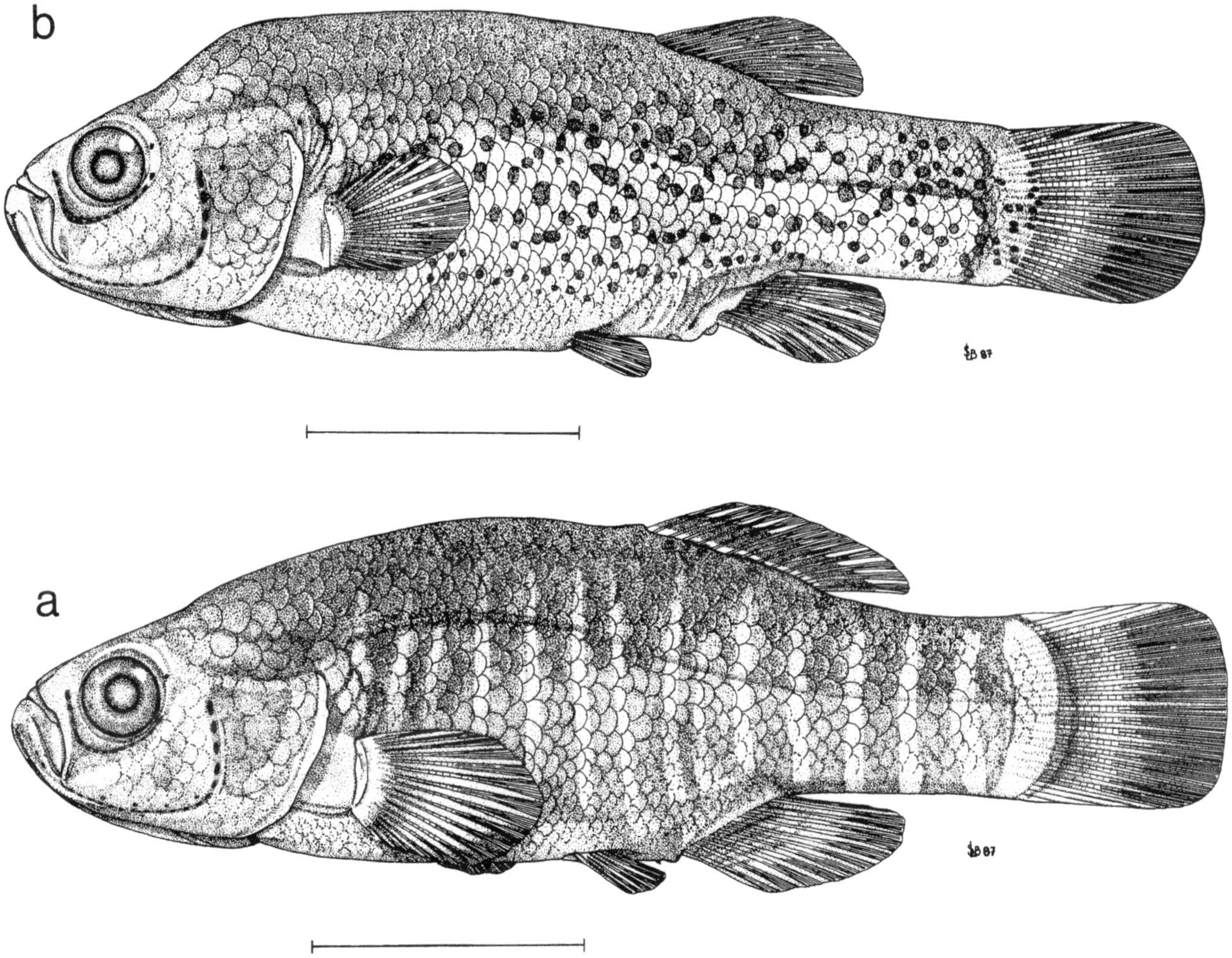

Fig. 1. Male (*a*) holotype of *Aphanius vladykovi* (NMC 79-0247), 36.6 mm SL and adult (*b*) female (NMC 79-0247A, 36.0 mm SL.

Description

The general shape of a male and a female *A. vladykovi* can be seen in Figure 1. Table 1 summarises meristic characters and Table 2 morphometric characters in *A. vladykovi,* and in *A. sophiae* the putative closest relative. Table 3 summarises differences between sexes within both these species and between sexes between both species. Eight of 10 meristic characters are significantly different between the two species and these apply equally to both sexes except for vertebral characters. Significant differences are found in 13 of 21 morphometric characters and these apply equally to both sexes except for the caudal peduncle character. There were no differences within *A. vladykovi* (or *A. sophiae*) between sexes for meristic characters. Sexual dimorphism in *A. vladykovi* was found in 10 of 21 morphometric characters and this was matched by *A. sophiae* with one addition (pelvic fin length in head length).

Scale rows around the caudal peduncle numbered 20–25 in *A. vladykovi* and 15–20 in *A. sophiae,* scales above the lateral line were 6–9 and 4–7 respectively, scales between the lateral line and the anal fin were 8–12 and 5–8 respectively and scales between the lateral line and the pelvic fin were 11–15 and 6–9 respectively. These figures indicate that scales are smaller and more numerous all over the body in the new species. Scales are regularly

Table 1. Meristic characters of *Aphanius vladykovi* and *A. sophiae* (n = number of specimens, x = mean, S.D. = standard deviation; underlined numbers = value for holotype).

Lateral line scales

		25	26	27	28	29	30	31	32	33	34	35	36	37	38	39	40	41	42	43	44	45	46	47	n	x	S.D.
A. vladykovi	♂	–	–	–	–	–	–	–	–	–	–	–	1	5	12	4	$\underline{6}$	4	3	–	–	–	1	1	37	39.4	2.35
	♀	–	–	–	–	–	–	–	–	–	–	–	1	3	3	2	4	2	–	1	–	–	–	–	16	39.0	1.86
A. sophiae	♂	–	–	6	9	7	7	1	–	–	–	–	–	–	–	–	–	–	–	–	–	–	–	–	30	28.6	1.16
	♀	1	–	1	5	10	6	–	–	–	–	–	–	–	–	–	–	–	–	–	–	–	–	–	23	28.8	1.17

Total dorsal fin rays

		11	12	13	14	15	n	x	S.D.
A. vladykovi	♂	1	17	$\underline{18}$	1	–	37	12.5	0.61
	♀	3	4	9	–	–	16	12.4	0.81
A. sophiae	♂	–	7	13	9	1	30	13.1	0.82
	♀	1	3	14	4	1	23	13.0	0.83

Total anal fin rays

		10	11	12	13	n	x	S.D.
A. vladykovi	♂	–	8	$\underline{18}$	11	37	12.1	0.72
	♀	–	5	7	4	16	11.9	0.77
A. sophiae	♂	1	11	12	6	30	11.8	0.82
	♀	–	7	15	1	23	11.7	0.54

Total pectoral fin rays

		14	15	16	17	18	19	n	x	S.D.
A. vladykovi	♂	2	12	18	$\underline{5}$	–	–	37	15.7	0.78
	♀	2	6	7	1	–	–	16	15.4	0.81
A. sophiae	♂	1	4	18	5	2	–	30	16.1	0.85
	♀	–	3	13	5	1	1	23	16.5	0.93

Total pelvic fin rays

		5	6	7	8	n	x	S.D.
A. vladykovi	♂	6	$\underline{30}$	–	1	37	5.9	0.52
	♀	3	12	1	–	16	5.9	0.50
A. sophiae	♂	4	26	–	–	30	5.9	0.35
	♀	5	18	–	–	23	5.8	0.42

Total gill rakers

		10	11	12	13	n	x	S.D.
A. vladykovi	♂	7	$\underline{27}$	3	–	37	10.9	0.52
	♀	1	14	1	–	16	11.0	0.37
A. sophiae	♂	1	18	9	2	30	11.4	0.68
	♀	1	20	12	–	23	11.5	0.59

Total vertebrae

		27	28	29	30	n	x	S.D.
A. vladykovi	♂	2	15	$\underline{19}$	1	37	28.5	0.65
	♀	–	9	7	–	16	28.4	0.51
A. sophiae	♂	3	21	6	–	30	28.1	0.55
	♀	5	8	10	–	23	28.2	0.80

Precaudal vertebrae

		10	11	12	13	n	x	S.D.
A. vladykovi	♂	1	5	$\underline{23}$	8	37	12.0	0.69
	♀	–	–	13	3	16	12.2	0.40
A. sophiae	♂	–	7	19	4	30	11.9	0.60
	♀	–	4	18	1	23	11.9	0.46

Table 1. Continued.

		Caudal vertebrae 15	16	17	n	x	S.D.
A. vladykovi	♂	1	17	19	37	16.5	0.56
	♀	–	12	4	16	16.3	0.45
A. sophiae	♂	2	20	8	30	16.2	0.55
	♀	1	13	9	23	16.3	0.57

	Male flank bars 9	10	11	12	13	14	15	16	17	18	19	20	21	n	x	S.D.
A. vladykovi	3	8	8	10	3	5	–	–	–	–	–	–	–	37	11.5	1.48
A. sophiae	–	4	1	2	3	5	4	6	3	1	–	–	1	30	14.3	2.63

Table 2. Morphometrics for *Aphanius vladykovi* and *A. sophiae* (values for male holotype included in total male values).

			Holotype	Number	Range	Mean	Standard deviation
Standard length / Head length	*A. vladykovi*	♂	3.1	37	3.0–3.5	3.2	0.10
		♀		16	3.0–3.5	3.3	0.13
	A. sophiae	♂		30	3.2–3.6	3.5	0.11
		♀		23	3.2–4.0	3.6	0.18
Standard length / Predorsal length	*A. vladykovi*	♂	1.5	37	1.5–1.6	1.5	0.04
		♀		16	1.5–1.6	1.5	0.05
	A. sophiae	♂		30	1.6–1.8	1.7	0.05
		♀		23	1.6–1.7	1.7	0.05
Standard length / Prepelvic length	*A. vladykovi*	♂	1.8	37	1.7–1.9	1.8	0.05
		♀		16	1.7–1.9	1.8	0.05
	A. sophiae	♂		30	1.8–2.0	1.9	0.07
		♀		23	1.7–2.0	1.9	0.09
Standard length / Preanal length	*A. vladykovi*	♂	1.5	37	1.4–1.6	1.5	0.02
		♀		16	1.4–1.5	1.5	0.04
	A. sophiae	♂		30	1.5–1.6	1.5	0.04
		♀		23	1.4–1.5	1.5	0.04
Standard length / Body depth	*A. vladykovi*	♂	3.0	37	2.8–3.4	3.0	0.16
		♀		16	3.0–3.3	3.1	0.11
	A. sophiae	♂		30	2.7–3.9	3.2	0.25
		♀		23	3.0–3.9	3.4	0.23
Standard length / Head depth	*A. vladykovi*	♂	3.8	37	3.5–4.3	3.9	0.16
		♀		16	3.7–4.2	4.0	0.14
	A. sophiae	♂		30	3.7–4.4	4.1	0.18
		♀		23	3.9–4.7	4.3	0.23
Head length / Head width	*A. vladykovi*	♂	1.6	37	1.4–1.6	1.5	0.07
		♀		16	1.4–1.5	1.5	0.05
	A. sophiae	♂		30	1.4–1.7	1.5	0.07
		♀		23	1.2–1.7	1.5	0.11

Table 2. (Continued).

			Holotype	Number	Range	Mean	Standard deviation
Head length / Head depth	*A. vladykovi*	♂	1.2	37	1.1–1.3	1.2	0.05
		♀		16	1.2–1.3	1.2	0.04
	A. sophiae	♂		30	1.1–1.3	1.2	0.05
		♀		23	1.1–1.3	1.2	0.07
Head length / Orbit diameter	*A. vladykovi*	♂	3.6	37	2.7–3.6	3.2	0.22
		♀		16	2.8–3.9	3.2	0.29
	A. sophiae	♂		30	2.9–3.6	3.2	0.18
		♀		16	2.8–3.6	3.2	0.19
Head length / Snout length	*A. vladykovi*	♂	3.6	37	3.4–4.2	3.9	0.23
		♀		16	3.3–4.1	3.9	0.23
	A. sophiae	♂		30	3.6–4.2	3.9	0.18
		♀		16	3.2–4.2	3.8	0.26
Head length / Interorbital width	*A. vladykovi*	♂	2.5	37	2.2–2.8	2.5	0.13
		♀		16	2.3–2.8	2.5	0.15
	A. sophiae	♂		30	2.2–2.7	2.5	0.14
		♀		23	2.3–2.7	2.5	0.13
Head length / Postorbital length	*A. vladykovi*	♂	2.0	37	1.9–2.3	2.1	0.10
		♀		16	1.9–2.3	2.1	0.12
	A. sophiae	♂		30	2.0–2.4	2.2	0.11
		♀		23	2.0–2.3	2.2	0.08
Head length / Mouth width	*A. vladykovi*	♂	3.2	37	2.7–3.6	3.2	0.21
		♀		16	2.8–3.4	3.1	0.19
	A. sophiae	♂		30	2.8–3.6	3.2	0.19
		♀		23	2.8–3.5	3.2	0.19
Head length / Dorsal fin length	*A. vladykovi*	♂	1.2	37	1.1–1.4	1.2	0.07
		♀		16	1.3–1.6	1.4	0.09
	A. sophiae	♂		30	0.9–1.2	1.0	0.08
		♀		23	1.0–1.4	1.2	0.12
Head length / Anal fin length	*A. vladykovi*	♂	1.5	37	1.3–1.7	1.5	0.08
		♀		16	1.4–1.8	1.6	0.12
	A. sophiae	♂		30	1.1–1.5	1.3	0.09
		♀		23	1.3–1.6	1.4	0.12
Head length / Pectoral fin length	*A. vladykovi*	♂	1.7	37	1.6–1.9	1.8	0.09
		♀		16	1.8–2.3	1.9	0.13
	A. sophiae	♂		30	1.5–1.8	1.6	0.09
		♀		23	1.4–2.0	1.7	0.08
Head length / Pelvic fin length	*A. vladykovi*	♂	4.1	37	3.5–5.5	4.2	0.42
		♀		16	3.3–5.5	4.3	0.58
	A. sophiae	♂		30	2.5–3.5	2.9	0.24
		♀		23	2.7–3.7	3.2	0.34
Dorsal fin length / Anal fin length	*A. vladykovi*	♂	1.3	37	1.1–1.3	1.2	0.07
		♀		16	1.0–1.3	1.1	0.09
	A. sophiae	♂		30	1.1–1.4	1.3	0.08
		♀		23	1.0–1.4	1.2	0.11

Table 2. (Continued).

			Holotype	Number	Range	Mean	Standard deviation
Pectoral-pelvic fin distance / Pectoral fin length	*A. vladykovi*	♂	1.3	37	1.1–1.5	1.3	0.09
		♀		16	1.4–2.0	1.6	0.14
	A. sophiae	♂		30	1.1–1.6	1.3	0.09
		♀		23	1.1–2.0	1.5	0.21
Pelvic-anal fin distance / Pelvic fin length	*A. vladykovi*	♂	1.3	37	1.3–2.4	1.7	0.24
		♀		16	1.5–2.8	2.0	0.32
	A. sophiae	♂		30	1.1–1.7	1.4	0.15
		♀		23	1.3–2.0	1.6	0.14
Caudal peduncle length / Caudal peduncle depth	*A. vladykovi*	♂	1.3	37	1.2–1.5	1.3	0.09
		♀		16	1.4–1.6	1.4	0.07
	A. sophiae	♂		30	1.2–1.6	1.3	0.13
		♀		23	1.4–1.8	1.5	0.11

Table 3. Meristic and morphometric differences within and between male and female *A. vladykovi* and *A. sophiae* (p at left is significance probability for differences between male and female *A. vladykovi;* p at right is the same for *A. sophiae;* central p values are between species for the same sex).

Character		*A. vladykovi*		*A. sophiae*
Lateral line scales	♂	$p>0.05$	$p<0.001$	$p>0.05$
	♀		$p<0.001$	
Total dorsal fin rays	♂	$p>0.05$	$p<0.01$	$p>0.05$
	♀		$p<0.05$	
Total anal fin rays	♂	$p>0.05$	$p>0.05$	$p>0.05$
	♀		$p>0.05$	
Total pectoral fin rays	♂	$p>0.05$	$p<0.05$	$p>0.05$
	♀		$p<0.01$	
Total pelvic fin rays	♂	$p>0.05$	$p>0.05$	$p>0.05$
	♀		$p>0.05$	
Total gill rakers	♂	$p>0.05$	$p<0.01$	$p>0.05$
	♀		$p<0.01$	
Total vertebrae	♂	$p>0.05$	$p<0.01$	$p>0.05$
	♀		$p>0.05$	
Precaudal vertebrae	♂	$p>0.05$	$p>0.05$	$p>0.05$
	♀		$p<0.05$	
Caudal vertebrae	♂	$p>0.05$	$p<0.05$	$p>0.05$
	♀		$p>0.05$	
Male flank bars			$p<0.001$	
Standard length / Head length	♂	$p<0.01$	$p<0.001$	$p<0.02$
	♀		$p<0.001$	
Standard length / Predorsal length	♂	$p>0.05$	$p<0.001$	$p>0.05$
	♀		$p<0.001$	
Standard length / Prepelvic length	♂	$p>0.05$	$p<0.001$	$p>0.05$
	♀		$p<0.001$	
Standard length / Preanal length	♂	$p>0.05$	$p>0.05$	$p>0.05$
	♀		$p>0.05$	
Standard length / Body depth	♂	$p<0.05$	$p<0.001$	$p<0.01$
	♀		$p<0.001$	

Table 3. (Continued).

Character		*A. vladykovi*	*A. sophiae*	
Standard length	♂	p<0.05	p<0.001	p<0.001
Head depth	♀		p<0.001	
Head length	♂	p>0.05	p>0.05	p>0.05
Head width	♀		p>0.05	
Head length	♂	p>0.05	p>0.05	p>0.05
Head depth	♀		p>0.05	
Head length	♂	p>0.05	p>0.05	p>0.05
Orbit diameter	♀		p>0.05	
Head length	♂	p>0.05	p>0.05	p>0.05
Snout length	♀		p>0.05	
Head length	♂	p>0.05	p>0.05	p>0.05
Interorbital width	♀		p>0.05	
Head length	♂	p>0.05	p<0.001	p>0.05
Postorbital length	♀		p<0.02	
Head length	♂	p>0.05	p>0.05	p>0.05
Mouth width	♀		p>0.05	
Head length	♂	p<0.001	p<0.001	p<0.001
Dorsal fin length	♀		p<0.001	
Head length	♂	p<0.001	p<0.001	p<0.001
Anal fin length	♀		p<0.001	
Head length	♂	p<0.01	p<0.001	p<0.01
Pectoral fin length	♀		p<0.001	
Head length	♂	p>0.05	p<0.001	p<0.001
Pelvic fin length	♀		p<0.001	
Dorsal fin length	♂	p<0.001	p<0.001	p<0.001
Anal fin length	♀		p<0.01	
Pectoral-pelvic fin distance	♂	p<0.001	p>0.05	p<0.001
Pectoral fin length	♀		p>0.05	
Pelvic-anal fin distance	♂	p<0.001	p<0.001	p<0.001
Pelvic fin length	♀		p<0.001	
Caudal peduncle length	♂	p<0.001	p>0.05	p<0.001
Caudal peduncle depth	♀		p<0.01	

arranged over the body, embedded and each scale overlaps adjacent scales. Anterior flank scales are a vertical oval with circuli and radii restricted to the posterior field. Numbers of circuli and radii are body and scale size dependent and so are fewer in *A. vladykovi* than in *A. sophiae* of comparable body size. Teeth are found in both jaws of the superior mouth in a single row and are tricuspid. Teeth are hyaline or discoloured with brown in patches. Detailed tooth counts were not made as this character varies widely and may be size dependent. There are up to 21 teeth in the upper jaw of *A. vladykovi.* Gill rakers are short, just reaching the base of the adjacent raker when adpressed to the gill arch. All fins are rounded distally. The origin of the anal fin is enclosed by a fleshy sheath which is most apparent in large females but is also present in males. The peritoneum is dark brown and the gut has a single large loop.

The cephalic sensory pores are reduced to a series of neuromasts in both *A. vladykovi* and *A. sophiae.* Live adult males bear creamy bars on yellowish flanks. The pelvic fins and the edge of the caudal fin are yellow, the anal fin has an orange-yellow edge and the pectoral fins are orange-yellow. The dorsal fin is white with a wide blue-black band in the centre and a narrow blue-black band at the base. In some males the basal band is absent.

Live adult females have a slight bluish tinge to their flanks and flank spots are brown. Pigmentation in preserved specimens is as follows. The dorsal fin bands are dark and remain darker in preservative than the pigmentation seen in male *A. sophiae* which is more diffuse, tending to several horizontal rows of speckles, and which is not formed into a strongly distinct band or bands. Other fins are lightly pigmented being finely speckled with melanophores with the dorsal and pectoral fins the darkest. The anal fin is fleshy and not hyaline but pigmentation is light and not dark as in *A. sophiae*. There is some tendency to formation of one or two narrow bars on the anal fin, paralleling the base, but these are weakly developed or absent in most fish. A mid-dorsal stripe is not apparent, perhaps because the dorsal flanks and the back are darker than in females. Flank bars are light alternating with darker and broader bars. The lighter bars become broader posteriorly. There is a light, wide bar at the base of the tail which was not included in counts in Table 1. The sides of the head and upper and lower jaws are pigmented with scattered melanophores while the underside of the head is not. The belly is also free of extensive melanophores. The most characteristic feature of females is the large number of scattered spots on the flank, extending from behind the head to the base of the tail. These spots do not extend to the dorsal flank or the back. The distinctive, large and often lozenge-shaped spot on the central caudal base of *A. sophiae* females is absent from the new species. The upper flank and the back are more heavily pigmented than the flanks (except for the flank spots). The lower surface of the head and the belly are melanophore free as in males and the sides of the head and the upper and lower jaws are heavily pigmented. The dorsal fin is finely pigmented on the rays and membranes and is generally darker than the caudal fin which is also finely pigmented. The other fins have very little pigment but all are fleshy and therefore not hyaline. There is some suggestion of a thin horizontal stripe near the base of the dorsal fin in the largest specimens but this is never well developed.

Zoogeography

Fosssil cyprinodonts have been described from the Oligocene of the Paris basin and the early Tertiary of the Aral Sea when this vast area was occupied by the Tethys Sea (Kosswig 1967, Parenti 1981) and Priem (1908) has described a cyprinodont from the Miocene of Iran. Kosswig (1955) regards *Aphanius* in the Middle East as relicts of the Tethys. The mountain populations of *A. vladykovi* and *A. sophiae* in Iran may well have risen with the post-Pliocene uplift of the Zagros Mountains. *Aphanius* species are small fishes with a low dispersal ability (Kosswig 1967) and it is here considered unlikely that the ancestor of *A. vladykovi* dispersed from lowland populations into an existing, high range of mountains represented today by the Zagros.

Oberlander (1965) has carried out an extensive study on the origin of the Zagros drainage patterns. Streams do not follow the line of least resistance but often incise the higher part of mountain ridges which are anticlines composed of the Oligocene-Miocene Asmari Limestone. This situation arises because the rising Zagros ridges were eroded by a superposed drainage pattern already antecedent to the exhumation of overlying sediments of later Miocene marls and evaporites. The trend of the Zagros ridges is in a NW-SE direction and streams often cross this direction at right angles. This has resulted in the development of *tangs*, narrow clefts in the Zagros ridges with very swift waters, which may be up to 2400 metres deep with vertical sides rising from the water surface to over 300 metres. *A. vladykovi* is found in the uppermost reaches of the Karun River basin which drains to the head of the Persian Gulf and is a tributary of the Tigris-Euphrates basin. In this locality it is isolated above a series of *tangs* which effectively block gene flow between *A. vladykovi* and lowland populations of other *Aphanius* species including those referred to *A. sophiae*.

Relationships

The closest relative of *A. vladykovi* is apparently *A. sophiae*. This conclusion is based on preliminary

Fig. 2. Type locality of *Aphanius vladykovi.*

studies of a number of populations of *Aphanius* species in Iran and because of the observations of Parenti (1981). She has distinguished derived members of the genus *Aphanius* as *'Aphanius'*, without formally designating a new genus. One of the distinguishing features of *'Aphanius'* is that cephalic sensory pores are reduced to neuromasts, a character observed by Parenti for *A. mento,* which is also found in *A. vladykovi* and *A. sophiae*. Particular differences are found between these three species in pigmentation patterns. *A. vladykovi* and *A. sophiae* are distinct but the presence of male flank bars and female flank spots are shared between the two species in contrast to other Iranian *Aphanius*. *A. mento* males have silvery-speckled fins and body (see Goren & Rychwalski 1978, Coad 1980, Krupp 1983 and Villwock et al. 1983 for illustrations and descriptions).

The high number of scales found in *A. vladykovi* is only matched by some populations of species in the genus *Anatolichthys* Kosswig and Sozer, 1945. This genus is restricted to four localities in southwestern Anatolia (Kosswig 1964) and has been synonymised with *Kosswigichthys* Sozer, 1942 (Parenti 1981) or *Aphanius* Nardo, 1827 (Villwock 1982). *'Anatolichthys'* populations are polymorphic for scale counts which reach as high as 55 through reduction in scale size. Some high scale count populations are in unusually bitter, salty lakes to which they have, perhaps, become adapted, although the adaptation is believed to be physiological with high scale counts incidental (Kosswig 1961, Villwock 1964, 1970). The high scale count of *A. vladykovi* is not due to an unusual reduction in scale size and this new species does not show loss of scales and an individual variability in scale size which are remarkable features of *'Anatolichthys'* (see Villwock 1964: Fig. 8, 1984: Fig. 1, 10). The new species from Iran is isolated from other populations of *Aphanius* sensu lato, particularly those of southwest Anatolia, and is not in water of an unusual chemical composition.

Habitat

The type locality lies at an altitude of ~2380 metres on a small plain between high ranges at the easternmost edge of the Zagros Mountains (Fig. 2). The Kuh-e Kukalar range to the southwest reaches 4298 metres. The type locality was a large pool, artificially dammed, in the upper reaches of the Karun River basin which drains to the head of the Persian Gulf. Pool width was estimated at 300 metres and specimens were captured in shallow water at about 40 cm depth. The pool was composed of fresh water which was cloudy over a bottom of mud and pebbles. The shore was grassy and the pool contained large amounts of *Myriophyllum* and marginal rushes. At 1515 hours on a warm, windy and sunny day, water temperature was 29° C, pH was 6.5 and conductivity 0.2 millimhos. The second locality was slightly lower in altitude and was a mud-bottomed stream 3–4 metres wide with pools up to 1 metre deep. At 1545 hours water temperature was 22° C, pH was 6.2 and conductivity was 0.45 millimhos. Current was slow and the stream had moderate amounts of *Myriophyllum*. The stream banks lacked bushes or trees.

Etymology

The new species is named for the late Professor V.D. Vladykov who worked in Iran on the fisheries of the Caspian Sea basin, collected extensive material there and accepted the author as his graduate student to study those specimens.

Acknowledgements

I am indebted to Dr. F. Krupp, Forschungsinstitut Senckenberg, Frankfurt, Dr. Don E. McAllister and Claude B. Renaud, Ichthyology Section, National Museum of Natural Sciences, Ottawa for their comments on this manuscript.

References cited

Berg, L.S. 1949. Freshwater fishes of Iran and adjacent countries. Trudy Zool. Inst. Akad. Nauk SSSR 8: 783–858.

Coad, B.W. 1980. A re-description of *Aphanius ginaonis* (Holly, 1929) from southern Iran (Osteichthyes: Cyprinodontiformes). J. Nat. Hist. 14: 33–40.

Coad, B.W. 1987. Zoogeography of the freshwater fishes of Iran. Beihefte zum Tübinger Atlas des Vorderen Orients, Reihe A (Naturwissenschaften), Wiesbaden, 28: 213–228.

Goren, M. & E.M. Rychwalski. 1978. Hybrids of *Aphanius dispar* and *Aphanius mento* (Cyprinodontidae: Pisces). Zool. J. Linn. Soc. 63: 259–264.

Hubbs, C.L. & K.F. Lagler. 1958. Fishes of the Great Lakes Region. University of Michigan Press, Ann Arbor. xv + 213 pp.

Jenkins, J.T. 1910. Notes on fish from India and Persia, with descriptions of new species. I. – On a collection of fishes made by W.T. Blanford in 1872 in Persia and Baluchistan. Rec. Indian Mus. 5: 123–128.

Kosswig, C. 1955. Zoogeography of the Near East. Syst. Zool. 4: 49–73, 96.

Kosswig, C. 1961. Speciation in the earlier Central Anatolian Lake Basin. pp. 561–593. *In:* W.F. Blair (ed.) Vertebrate Speciation, A University of Texas Symposium, University of Texas Press, Austin.

Kosswig, C. 1964. Problems of polymorphism in fishes. Copeia 1964: 65–75.

Kosswig, C. 1967. Tethys and its relation to the peri-Mediterranean faunas of fresh-water fishes. pp. 313–324. *In:* C.G. Adams & D.V. Ager (ed.) Aspects of Tethyan Biogeography, Systematics Association Publication.

Krupp, F. 1983. Freshwater fishes of Saudi Arabia and adjacent regions of the Arabian Peninsula. Fauna of Saudi Arabia 5: 568–636.

Oberlander, T. 1965. The Zagros Streams. A new interpretation of transverse drainage in an orogenic zone. Syracuse Geographical Series 1: xii + 168 pp.

Parenti, L.R. 1981. A phylogenetic and biogeographic analysis of cyprinodontiform fishes (Teleostei, Atherinomorpha). Bull. Amer. Mus. Nat. Hist. 168: 335–557.

Priem, F. 1908. Poissons fossiles de Perse. pp. 1–25. *In:* J. de Morgan, Delegation en Perse, Ann. Hist. Nat. 1.

Tortonese, E. 1986. Cyprinodontidae. pp. 623–626. *In:* P.J.P. Whitehead, M.-L. Bauchot, J.-C. Hureau, J. Nielsen & E. Tortonese (ed.) Fishes of the North-eastern Atlantic and the Mediterranean, Vol. 2, Unesco, Paris.

Villwock, W. 1964. Genetische Untersuchungen an altweltlichen Zahnkarpfen der Tribus *Aphaniini* (*Pisces, Cyprinodontidae*) nach Gesichtspunkten der Neuen Systematik. Z. zool. Syst. Evol. 2: 267–382.

Villwock, W. 1970. Distribution, ecology and relationship of Near East and Mediterranean Cyprinodonts of the Genus *Aphanius*. Journées ichthyol., Rome, C.I.E.S.M.: 89–92.

Villwock, W. 1982. Aphanius (Nardo, 1827) and Cyprinodon (Lac., 1803) (Pisces: Cyprinodontidae), an attempt for a genetic interpretation of speciation. Z. zool. Syst. Evol. 20: 187–197.

Villwock, W. 1984. Curt Kosswigs ichthyologische Forschungen – ein Beitrag zur Faunengeschichte Anatoliens und zu Phänomenen der Evolution am Beispiel anatolischer Zahnkarpfen (Pisces: Cyprinodontidae). Mitt. hamb. zool. Mus. Int. 80: 19–40.

Villwock, W., A. Scholl & F. Krupp. 1983. Zur Taxonomie, Verbreitung und Speziation des Formenkreises *Aphanius dispar* (Rüppell, 1828) und Beschreibung von *Aphanius sirhani* n. sp. (Pisces: Cyprinodontidae). Mitt. hamb. zool. Mus. Inst. 80: 251–277.

Vadim had a keen eye for taxonomic characters and women! At the 1974 ASIH meeting, sponsored by the National Museum of Natural Sciences, Ottawa, V.D. Vladykov to the right of Brian and Sylvie Coad.

Environmental Biology of Fishes Vol. 23, No. 1–2, pp. 127–130, 1988

Lampetra (*Eudontomyzon*) *gracilis*, a synonym of *Eudontomyzon danfordi*

Claude B. Renaud[1] & Juraj Holčík[2]
[1] *Department of Biology, University of Ottawa, Ottawa, Ontario K1N 6N5, Canada and Ichthyology Section, National Museum of Natural Sciences, Ottawa, Ontario K1P 6P4, Canada*
[2] *Ústav rybárstva a hydrobiológie, Drieňová ul.č.3, 826 24 Bratislava, Czechoslovakia*

Received 23.12.1987 Accepted 5.1.1988

Key words: *Lamprey taxonomy, Czechoslovakia, Tisza River, Parasitic, Nonparasitic, Agnatha, Petromyzontidae*

Synopsis

Evidence is presented which suggests that the nonparasitic lamprey, *Lampetra* (*Eudontomyzon*) *gracilis* Kux, 1965, is conspecific with the parasitic lamprey *Eudontomyzon danfordi* Regan, 1911. The diagnostic characters of the holotype and of the non-type material of *E. gracilis* are features found in *E. danfordi* specimens in their second and final year of adult life, thereby making the former a junior synonym of the latter.

Introduction

Kux (1965) described a new species of nonparasitic lamprey, *Lampetra* (*Eudontomyzon*)[1] *gracilis*, endemic to the Tisza River basin (Toplá River, type-locality; Hornád and Okna rivers; and Ulička Rivulet) in eastern Slovakia. A number of authors (Bănărescu 1969, Holčík 1970, Holčík & Hensel 1972, Renaud 1982) have expressed doubts as to the specific distinctiveness of *E. gracilis*. According to Kux (1965, 1967, 1969, 1974), the distribution of *E. gracilis* is almost completely sympatric with that of the parasitic species *E. danfordi* Regan, 1911. He maintains that both species are identical in external morphology of ammocoetes and adults, dentition of the oral disc, duration of the larval period, pigmentation of ammocoetes, timing and duration of metamorphosis, and type of habitat. The differences between the two species, according to Kux (1965), are in their life span and trophic ecology after metamorphosis. *E. danfordi* lives approximately 21 months as an adult, feeding actively in the summer following metamorphosis and spawning in the spring after that, while *E. gracilis* lives only 8 to 9 months as an adult, does not feed at all during adulthood and spawns in the spring following metamorphosis. Kux (1965) adds that the two can be distinguished between the 2nd or 3rd month to the 9th month post metamorphosis, only on the basis of differences in intestinal and gonadal development. During this period, the intestinal diameter in *E. danfordi* is large (about 2 mm) and the gonads poorly developed, while in *E. gracilis* the intestinal diameter is narrow (0.7–1.5 mm) and the gonads well developed. Once the adult trophic phase is over in *E. danfordi*, the intestine and gonads become as in *E. gracilis*, making the two species indistinguishable once again.

[1] We follow the majority of authors (see Renaud 1982) in treating *Eudontomyzon* as a distinct genus rather than a subgenus of *Lampetra*.

This study demonstrates that *E. gracilis* is conspecific with and a junior synonym of *E. danfordi*.

Material

Five metamorphosed specimens, identified by Kux as *Lampetra* (*Eudontomyzon*) *gracilis,* were examined. Of these, three came from the Hornád River basin, the only river basin reported by Kux (1965, 1969) to be inhabited solely by *E. gracilis*. The collection data for the above specimens are as follows:

Tag W (VDV)5817; ♀; 158 mm TL; Ulička Rivulet, Ulič, Slovakia; 48°58′N 22°26′E; 16 Oct. 1962, T. Weisz.

Untagged; ♀; 140 mm TL; Toplá River, Slovakia; 49°18′N 21°5′E; 22 July 1964; collector unknown. This topotypic specimen was part of aquarium experiment no. 5 in Kux (1967, p. 213). It was collected as an ammocoete on 22 July 1964, it metamorphosed in August 1965, spawned on 15 May 1966, and subsequently died.

Untagged; 2 ♂: 1 ♀; 153–171 mm TL; Veľká Biela Voda Brook just above its confluence with the Hornád River, Slovakia; 48°53′N 20°26′E; 29 May 1975; A. Kirka.

Results and discussion

All nonparasitic species of Petromyzontidae except *E. gracilis* have a wingless velar apparatus in the adult period; wings having only been found in three parasitic species (Vladykov & Kott 1976). In a lamprey from Ulička Rivulet, identified by Kux as *E. gracilis*, and in three lampreys from the Hornád River basin, said by Kux (1965, 1969) to be inhabited exclusively by *E. gracilis*, wings with three and four tentacles were present on the velar apparatus. Vladykov et al. (1982) reported a range of two to four tentacles per wing in *E. danfordi*. The total number of velar tentacles in the aforementioned specimens was 10–13 ($\bar{x}$ = 11.3), which lies within the range of 9–15 for *E. danfordi* (Vladykov et al. 1982). Furthermore, within the genus *Eudontomyzon,* the presence of wings on the velar apparatus is a characteristic exclusive to individuals of parasitic *E. danfordi* (Vladykov & Kott 1976, Vladykov et al. 1982) that have begun their adult trophic phase (Renaud 1982, Renaud & Holčík 1986).

The identical larval life span of 4+ to 5+ years in *E. gracilis* and *E. danfordi* (Kux 1965), is at odds with data which show that the nonparasitic member of a paired lamprey species has a longer larval period than its parasitic counterpart (Hardisty & Potter 1971, Vladykov & Kott 1979, Hardisty 1979, 1986, Potter 1980); the nonparasitic lamprey replace the adult trophic phase by extending the microphagous larval life span. Kux (1965) described *E. gracilis* as a nonparasitic derivative of *E. danfordi*.

In his aquarium experiments, Kux (1967) marked the ammocoetes by making notches in their fins and many of them were under observation for three years. The ages of ammocoetes from the field however, were indirectly determined using length frequency distributions (Kux 1965, 1967) and these await validation. The counting of statolith bands, either unstained or stained with oxytetracycline, has so far only been used to validate ages in larvae of *Petromyzon marinus* (Volk 1986) and *Ichthyomyzon greeleyi* (Medland & Beamish 1987). This cautionary note is made so that arguments based on age of ammocoetes and presented in this paper are taken in their proper perspective.

As a rule, in the parasitic stem species, the maximum length of ammocoetes and newly metamorphosed individuals is shorter and that of spawning adults longer than in the corresponding stages in the nonparasitic satellite species (Vladykov & Kott 1978, 1979), thereby, forming an effective barrier to hybridization since optimal fertilization is ensured only through the pairing of males and females of similar lengths (Hardisty & Potter 1971). This is not the case with *E. gracilis* and *E. danfordi*. Kux (1965) states that in nature the size range of prespawning adults in both species is 160–245 mm and that they cannot be distinguished in this phase nor in the ammocoete period (Kux 1965, 1969) on the basis of length. Furthermore, as determined from rearing experiments in aquaria (Kux 1967), the size range in recently metamorphosed *E. gracilis* was 160–201 mm ($\bar{x}$ = 180.7; N = 14) and in *E.*

danfordi it was 158–205 mm ($\bar{x}$ = 183.3; N = 4). As well, in adults of *E. gracilis* measured just prior to spawning or immediately after their death, the size range was 131–180 mm ($\bar{x}$ = 156.9; N = 16) and in those of *E. danfordi*, it was 155–207 mm ($\bar{x}$ = 182.8; N = 4). The difference between *E. gracilis* and *E. danfordi* in the prespawning phase is not statistically significant at the chosen confidence level of $P<0.05$. However, it is significant ($0.025<P>0.01$) in the case of spawning specimens.

Johal & Oliva (1980) could not distinguish between ammocoetes of *E. danfordi* and *E. gracilis* and put them together in their key.

Kux also did not find morphological differences between the two aside from putative intestinal and gonadal differences.

We re-interpret Kux's data in the following manner:

In his field observations, Kux was comparing *E. danfordi* adults in their first year of adult life with some in their second and final year of adult life, which he called *E. gracilis* (Renaud 1982). The intestinal and gonadal development in *E. danfordi* during those periods are consistent with the above interpretation. Indeed, Zanandrea (1959) reported intestinal diameters ranging from 1.5 to 8.0 mm in recently metamorphosed and feeding (i.e. immature) *E. danfordi*, while Vladykov (1927) reported intestinal diameters from 0.3 to 1.0 mm in spawning (i.e. mature) *E. danfordi*. Mihail (1962) observed two forms of *E. danfordi* in the Valea Ierii Brook (Tisza River basin), Romania, in the spring following metamorphosis: one short, slender, and grey, with weakly developed gonads, and another longer, thicker, and brownish, with well developed gonads and concluded that *E. danfordi* lives two years as an adult. The former is in pre-trophic phase as *E. danfordi* begins feeding in the summer following metamorphosis.

In Kux's (1967) aquarium experiments, sexually mature nonfeeding individuals (57.1%) and immature feeding individuals (42.9%) developed from samples of ammocoetes collected in the Toplá and Okna rivers. Inadequate feeding conditions prevailed in those experiments (Kux 1967). The fact that 'the majority of ammocoetes died, metamorphosis was mostly delayed by 1–2 years, and most of the ammocoetes held for a longer period did not transform even after 6–7 years of larval life' (Kux 1967, p. 210; transl. by J.H.) attests to the severe conditions in the aquaria. Furthermore, in two aquarium-reared females, 184 and 207 mm TL from Okna River, and identified by Kux as *E. danfordi*, he counted 8630 and 3180 eggs ($\bar{x}$ = 5905), respectively (Kux 1967). This represents a decrease from the normal fecundity of the species, which is 7500 to 10300 eggs ($\bar{x}$ = 8900) (Hardisty 1964), and is another indication of the harsh conditions in the aquaria. We suggest that the sexually mature nonfeeding individuals, interpreted by Kux as belonging to a new nonparasitic species, *E. gracilis*, were derived from *E. danfordi* ammocoetes which because of the harsh aquarium conditions extended their microphagous larval life by 1–2 years, thereby removing the need for the parasitic phase, and thus, approached a neotenic state. As evidence for this interpretation, we examined a spawning female of *E. gracilis*, which was part of aquarium experiment no. 5 (Kux 1967). Its dorsal fins were low and well separated, and no cloacal fin-like fold had developed, contrary to the normal condition in a spawning female; a situation also reported by Hubbs (1971) for neotenic *Entosphenus lethophagus* in nature.

Kux's aquarium experiments may provide some insight into lamprey evolution since sexually mature lampreys belonging to a parasitic species were artificially induced without the need for the parasitic phase. However, until a nonparasitic lamprey belonging to the genus *Eudontomyzon* and distinguishable from *E. danfordi* in its second year of adult life is found in the Tisza River basin, we consider *E. gracilis* to be a junior synonym of the former.

Acknowledgements

Don E. McAllister and C.G. Gruchy, National Museum of Natural Sciences, Ottawa, and R.J. Beamish, Pacific Biological Station, Nanaimo, made some constructive comments on this paper.

References cited

Bănărescu, P. 1969. Cyclostomata şi Chondrichtyes (ciclostomi şi selacieni). Fau. Rep. Soc. Rom. 12(1): 5–104. (In Roumanian).

Hardisty, M.W. 1964. The fecundity of lampreys. Arch. Hydrobiol. 60: 340–57.

Hardisty, M.W. 1979. Biology of the cyclostomes. Chapman and Hall, London. 428 pp.

Hardisty, M.W. 1986. General introduction to lampreys. pp. 19–83. *In:* J. Holčík (ed.) The Freshwater Fishes of Europe, Volume 1/I, AULA-Verlag GmbH, Wiesbaden. 313 pp.

Hardisty, M.W. & I.C. Potter. 1971. Paired species. pp. 249–277. *In:* M.W. Hardisty & I.C. Potter (ed.) The Biology of Lampreys, Volume 1, Academic Press, London. 423 pp.

Holčík, J. 1970. On the occurrence of *Lampetra planeri* (Bloch, 1784) in the Poprad River (northern Slovakia) and notes to its taxonomy and ecology. Věstník čs. Spol. zool. 34: 22–32.

Holčík, J. & K. Hensel. 1972. Ichtyologická príručka. Obzor, Bratislava. 219 pp.

Hubbs, C.L. 1971. *Lampetra (Entosphenus) lethophaga,* new species, the nonparasitic derivative of the Pacific lamprey. Trans. San Diego Soc. Nat. Hist. 16: 125–63.

Johal, M.S. & O. Oliva. 1980. Key characters of larvae of two European lamprey species (Cyclostomata). Věst. čs. Společ. zool. 44: 197–212.

Kux, Z. 1965. *Lampetra gracilis,* a new nonparasitic species of lamprey from eastern Slovakia. Čas. Mor. musea 50: 293–302. (In Czech).

Kux, Z. 1967. A contribution to the bionomy of lampreys (Petromyzonidae) in the Tisza region of eastern Slovakia. Čas. Mor. musea 52: 209–216. (In Czech).

Kux, Z. 1969. A contribution to the distribution of lampreys (Petromyzonidae) in ČSSR. Čas. Mor. musea 54: 203–22. (In Czech).

Kux, Z. 1974. Die Variabilität der Unterkieferzähne bei den Neunaugen (Petromyzoniformes) aus dem Gebiet der ČSSR. Čas. Mor. musea 59: 211–218.

Medland, T.E. & F.W.H. Beamish. 1987. Age validation for the mountain brook lamprey, *Ichthyomyzon greeleyi.* Can. J. Fish. Aquat. Sci. 44: 901–904.

Mihail, N. 1962. Zur Biologie und Ökologie von *Eudontomyzon danfordi.* Zool. Anz. 168: 139–143.

Potter, I.C. 1980. The Petromyzoniformes with particular reference to paired species. Can. J. Fish. Aquat. Sci. 37: 1595–1615.

Regan, C.T. 1911. A synopsis of the marsipobranchs of the order Hyperoartii. Ann. Mag. Nat. Hist. 7: 193–204.

Renaud, C.B. 1982. Revision of the lamprey genus *Eudontomyzon* Regan, 1911. M.Sc. Thesis, University of Ottawa, Ottawa. 146 pp.

Renaud, C.B. & J. Holčík. 1986. *Eudontomyzon danfordi* Regan, 1911. pp. 146–164. *In:* J. Holčík (ed.) The Freshwater Fishes of Europe, Volume 1/I, AULA-Verlag GmbH, Wiesbaden. 313 pp.

Vladykov, V.D. 1927. Zur Frage über das Absterben unserer Lampreten (Petromyzonidae) nach der Laichzeit. Zool. Anz. 74: 237–239.

Vladykov, V.D. & E. Kott. 1976. The taxonomic significance of velar tentacles in Holarctic lampreys (Petromyzonidae). Rev. Trav. Inst. Pêches Marit. 40: 787–789.

Vladykov, V.D. & E. Kott. 1978. A new nonparasitic species of the Holarctic lamprey genus *Lenthenteron* Creaser and Hubbs, 1922, (Petromyzonidae) from northwestern North America with notes on other species of the same genus. Biol. Pap. Univ. Alaska (19): 1–74.

Vladykov,V.D. & E. Kott. 1979. Satellite species among the holarctic lampreys (Petromyzonidae). Can. J. Zool. 57: 860–867.

Vladykov, V.D., C.B. Renaud, E. Kott & P.S. Economidis. 1982. A new nonparasitic species of Holarctic lamprey, genus *Eudontomyzon* Regan 1911 (Petromyzontidae), from Greece. Can. J. Zool. 60: 2897–2915.

Volk, E.C. 1986. Use of calcareous otic elements (statoliths) to determine age of sea lamprey ammocoetes (*Petromyzon marinus*). Can. J. Fish. Aquat. Sci. 43: 718–722.

Zanandrea, G. 1959. Parasitic and nonparasitic lampreys of the Danube basin and the new systematic form *Eudontomyzon danfordi vladykovi.* Arch. Zool. Ital. 44: 215–250. (In Italian).

Environmental Biology of Fishes Vol. 23, No. 1–2, pp. 131–134, 1988

Shark attacks in the rivers of southern Iran

Brian W. Coad[1] & Frough Papahn[2]
[1] *Ichthyology Section, National Museum of Natural Sciences, P.O. Box 3443, Station D, Ottawa, Ontario, K1P 6P4, Canada*
[2] *Department of Biology, College of Science, Shahid Chamran University, Ahvaz, Khuzestan, Iran*

Received 30.11.1987 Accepted 15.12.1987

Key words: Elasmobranchs, Freshwater, Bull shark, *Carcharhinus*

Synopsis

The literature on shark attacks in freshwaters of southern Iran is reviewed and 11 attacks with 3 fatalities recorded from local informants for the period 1953 to 1985. The species of shark responsible is probably the bull shark, *Carcharhinus leucas,* whose presence in the Tigris River is confirmed by preserved specimens in the British Museum (Natural History). The Iranian records represent a significant proportion (~28%) of the documented cases world-wide for unprovoked, freshwater attacks.

Introduction

Records of sharks in rivers draining to the head of the Persian Gulf may be traced in the Arabic work 'Wonders of Creation' by Zakariya al-Qazwini published in 1263 A.D. and later translated into Persian. These fishes were reported from Basrah on the Tigris River and were formidable because of their voracity and teeth like the points of spears, swords or saws. Günther (1874) gave the first modern report of freshwater sharks, at Baghdad on the Tigris about 850 river km from the sea. Kennedy (1937), Khalaf (1961), Mahdi (1962) and Al-Daham (1976) confirm the presence of sharks far inland and Kennedy comments that, although less frequent at Baghdad than at Basrah nearer the mouth of the Tigris River, they were reported there every year.

Iranian reports of sharks are confirmed to brief mentions by Sykes (1902) and Blegvad & Loppenthin (1944) and to a more extensive article by Hunt (1951). Sykes (1902) saw sharks in the Ab-e-Gargar at Shushtar and said that men, women, children, horses and sheep were attacked and only the massive water buffalo was safe. Blegvad & Loppenthin (1944) noted that every year several people, especially children, fell victim to these sharks. Hunt (1951) was a member of the Royal Army Medical Corps, stationed at Ahvaz on the Karun River during the Second World War. He reported 13 cases of shark attack of which 12 died from their injuries. A further 10 cases were reported on from the Karun River, Khowr-e Bahmanshir and Shatt-al-Arab with 2 deaths. Many minor cases were probably not reported and the high mortality rate was attributed to the victims being mostly young children or old people either undernourished or feeble and to slow transport to medical facilities under the devastating summer heat on the Khuzestan plains. Size estimates for the sharks ranged up to about 2 metres and they were said to invade the Khowr-e Bahmanshir and the Karun River from July to September when freshwater flow was at a minimum and tidal penetration of salt water at its highest. Victims were paddling, swimming, bathing, washing vehicles or fishing. These records have been partly summarised in the International Shark Attack File which is now maintained at the National Underwater Accident Data Center, University of Rhode Island, Kingston, R.I. However this File has been without funds for gathering foreign data for many years and has no post-1960 freshwater records from Iran (John J. McAniff, in litt. 1986).

New records

In this note we report on attacks from 1953 to 1985. Information from local people indicates that attacks are common and occur every year although perhaps less frequently than in the past as direct use of the rivers for water supplies has declined. Local legend has it that attacks have decreased since shark oil is no longer used to caulk boats but this must rank with Blegvad & Loppenthin's (1944) report that sharks at Khorramshahr station gather under palm trees in order to eat the falling dates! Since the start of the Iran-Iraq war a number of soldiers have fallen victim to sharks but we have no details on these attacks.

The documented attacks are as follows. In April 1953 Mr. Nasser Seemrookh and Mr. Kaaby both lost an arm to sharks in the Karun River near Ahvaz but survived. Ahvaz is about 275 river km from the sea. The shark was estimated to be about 1.5 metres long and was grey in colour. Mr. Seemrookh was swimming in mid-river and the attack took place about 200 metres from a sewage outlet and about 400 metres from a slaughter-house. Mr. Seemrookh lost his right arm (and wrist-watch!) from the elbow downwards. In 1954 Mr. Kasem Jasem was attacked in the Karun River in the Hesamabad area of Shushtar and died shortly thereafter. Shushtar is about 420 river km from the sea. His son Abbas was attacked at the same place on the same day but survived to give this report. A girl was fatally attacked in a village a short distance from Hesamabad on the same day. These three incidents may be attributable to the same shark. In 1957 a Mr. Falah was attacked in Hesamabad but survived to report this incident. Also in 1957, a Mr. Paniry was attacked near Band Misan in Shushtar and survived. Mr. Kalantar and Mr. Nagib were attacked in Ahvaz in 1971 and survived. An undated report of an incident in Ahvaz resulted in the loss of a foot for Mr. Jabar-Kaaby. In 1985 a fisherman was attacked at Ramin near Ahvaz and was dead when pulled from the water. Attacks are also reported as a regular occurrence in modern times in the Khowr-e Bahmanshir. The victims were mostly swimming, their clothing was black, grey, white and in one case red, and attacks occurred in bright daylight, in moderate current and at relatively cool water temperatures. These rivers are turbid with very low water visibility. The circumstances surrounding each attack are fragmentary and no particular pattern can be discerned. A total of 11 attacks with 3 fatalities are verified for the period 1953–1985 and doubtless more occurred which were not reported or were not found in our local enquiries. The earlier records bring the total to 34 attacks with 17 fatalities. These notes indicate that shark attacks are a continuing problem in the rivers of southern Iran.

Discussion

The sharks are probably bull sharks, *Carcharhinus leucas* (Valenciennes in Müller and Henle, 1839). Garrick (1982) revised the genus *Carcharhinus* and examined the head of a specimen from Al Karradah near Baghdad which proved to be this species and one of us (BWC) has seen this specimen in the British Museum (Natural History), London (BM (NH)) where it bears the annotation that the fish was 1.25 metres long. A second whole specimen of this species, also from Baghdad, was 82 cm total length. Bull sharks are common in all tropical and subtropical seas and are known to travel far up rivers and into lakes including the Mississippi River, Lake Nicaragua and the Amazon River in the Americas, the Zambezi and Limpopo rivers of Africa, and the Hooghly river of India, Lake Jamoer of New Guinea and rivers and lakes of Australia. Compagno (1984) reports that bull sharks appear heavy and slow but are quick and agile when attacking prey. He believes that this species may well be the most dangerous tropical shark and

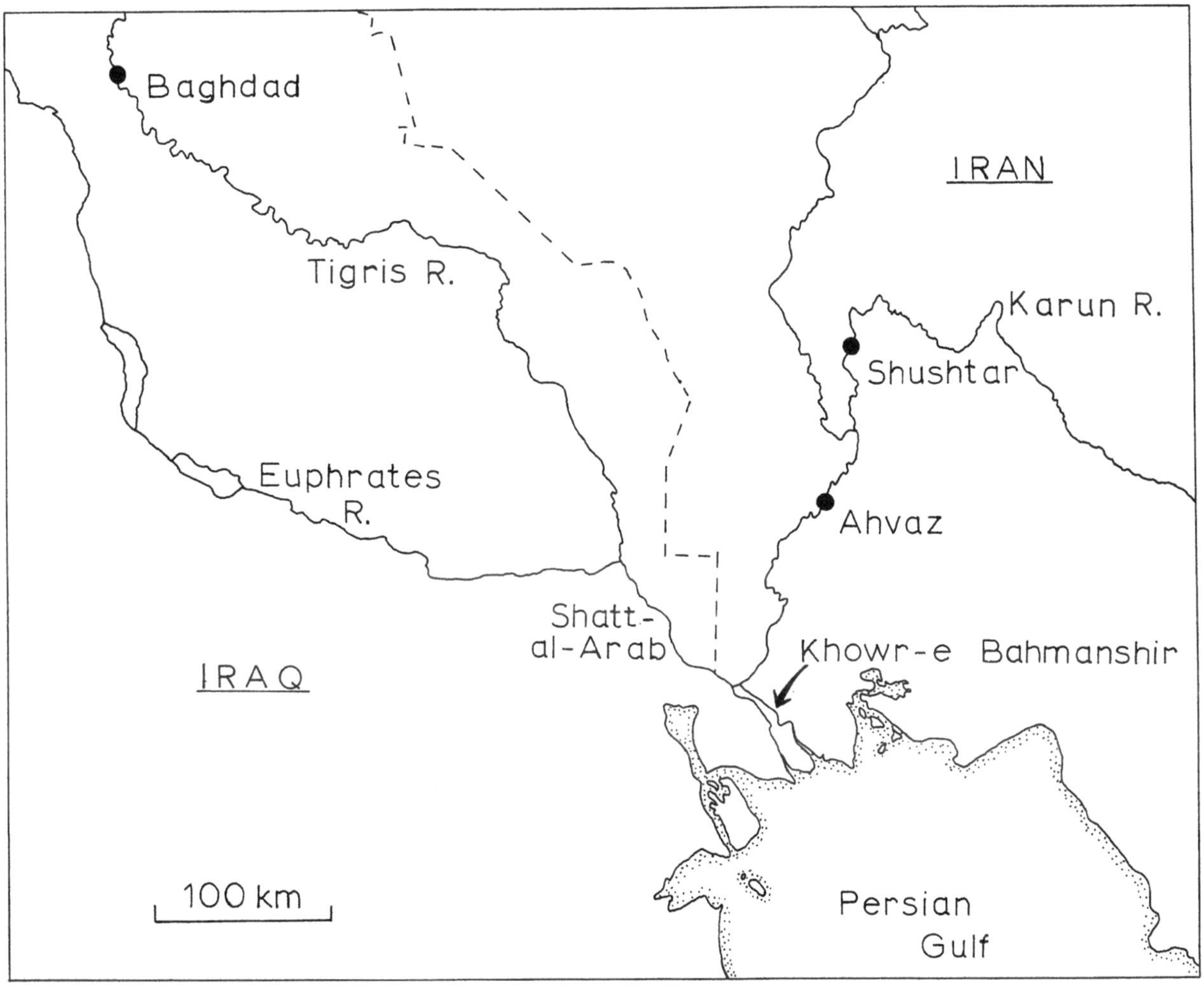

Fig. 1. Simplified drainage map of the rivers draining to the head of the Persian Gulf with shark attack localities.

the frequency of reported attacks in southern Iran bears this out.

Schultz and Malin in Gilbert (1963) summarise shark attacks for the world based on a documented file in the Division of Fishes of the United States National Museum. Unprovoked shark attacks in freshwater numbered less than 100 for the period 1837–1961 excluding Iranian records (a more accurate count is not possible because reports from several countries are vague as to numbers; 41 cases are from Australia where reports are best documented). Iranian records, numbering 34 in total from 1941–1985, are therefore a significant proportion (~28%) of reported freshwater attacks worldwide. In Iran half the attacks were fatal while for Australia about 60% were fatal.

Material examined

Carcharhinus leucas: BM(NH) 1874.4.28:9, Tigris River near Baghdad; BM(NH) 1924.10.1:1 (head only), Al Karradah on Tigris River near Baghdad.

Acknowledgements

We are grateful to our various informants who relived the unfortunate circumstances surrounding shark attacks on themselves and their relatives. BWC is grateful for the opportunity to examine material housed in the Fish Section of the British Museum (Natural History) through the hospitality of K. Banister and A. Wheeler. John J. McAniff,

Director, National Underwater Accident Data Center, University of Rhode Island consulted the International Shark Attack File for recent Iranian records.

References cited

Al-Daham, N.K. 1976. Fishes of Iraq and the Arab Gulf: Orders Squaliformes and Rajiformes. Bulletin of the Basrah Natural History Museum 3: 3–66.

Blegvad, H. & B. Loppenthin. 1944. Fishes of the Iranian Gulf. Danish Scientific Investigations in Iran, Part III. Einar Munksgaard, Copenhagen. 247 pp.

Compagno, L.J.V. 1984. Sharks of the world. An annotated and illustrated catalogue of shark species known to date. Part 2 – Carcharhiniformes. FAO Fisheries Synopsis 125, volume 4: 251–655.

Garrick, J.A.F. 1982. Sharks of the genus *Carcharhinus*. National Oceanic and Atmospheric Administration Technical Report, National Marine Fisheries Service Circular 445: 1–194.

Gilbert, P.W. (ed.). 1963. Sharks and survival. D.C. Heath, Boston. 578 pp.

Günther, A. 1874. A contribution to the fauna of the River Tigris. The Annals and Magazine of Natural History 4(14): 36–38.

Hunt, R.S. 1951. The sharks of Ahwaz. Journal of the Royal Army Medical Corps 97: 79–85.

Kennedy, W.P. 1937. Some additions to the fauna of Iraq. Journal of the Bombay Natural History Society 39: 745–749.

Khalaf, K.T. 1961. The marine and fresh water fishes of Iraq. Ar-Rabitta Press, Baghdad. 164 pp.

Mahdi, N. 1962. Fishes of Iraq. Ministry of Education, Baghdad. 82 pp.

Sykes, P.M. 1902. Ten thousand miles in Persia; or, eight years in Iran. John Murray, London. 481 pp.

Environmental Biology of Fishes Vol. 23, No. 1–2, pp. 135–139, 1988

Thermoregulatory activity in the Tecopa pupfish, *Cyprinodon nevadensis amargosae*, an inhabitant of a thermal spring

Robert W. McCauley[1] & Donald A. Thomson[2]
[1] *Department of Biology, Wilfrid Laurier University, Waterloo, Ontario N2L 3C5, Canada*
[2] *Department of Ecology and Evolutionary Biology, University of Arizona, Tucson, Arizona 85721, U.S.A.*

Received 15.12.1987 Accepted 16.1.1988

Key words: Temperature preference, Thermal tolerance, Warmwater fish, Optimum temperature, Tecopa, California, Cyprinodontiformes

Synopsis

The final thermal preferendum of Tecopa pupfish *Cyprinodon nevadensis amargosae* collected from a thermal spring was determined in the laboratory in spatial and temporal thermal gradients. Temperatures chosen centred around 30°C – a value close to those reported for other warmwater species in the literature. Upper limits to final preferenda and thermal tolerance in all species of fish are proposed.

Introduction

Populations of the pupfish, *Cyprinodon nevadensis* have been shown to be especially eurythermal. Its upper incipient lethal temperature is just below 40°C and it can survive winter temperatures as low as 3°C (Brown 1971, Otto & Gerking 1973). A subspecies, *C. nevadensis amargosae,* inhabits a thermal spring at Tecopa, California. This watercourse was created in 1967 when two artesian wells were dug. Geothermally heated water found its way into the Amargosa River and pupfish entered from the river through a connecting swamp to become residents of this new habitat (Deacon & Minckley 1974, Naiman 1976). The stream is a thermally challenging habitat since fish can encounter a wide range of temperature within a relatively short distance. In July, for instance, water at the source of the spring (Tecopa Bore) emerges at 47°C to cool to 30°C before entering the river (Brown 1971). Miller (1948) observed these fish in the river and reported that they were fairly common at temperatures of 33–36.5°C, with few individuals at 40°C. Brown (1971) observed that fish made brief excursions into water even above 40°C to graze on thermophyllic algae.

We predicted, therefore, that because of its flirtation with thermal death this pupfish would have, in addition to a high thermal tolerance, a well developed discrimination of water temperature. Its warm habitat suggested that its final preferendum (which often correlates with optimal growth temperature) would also be high. In general there is a direct relationship between upper lethal temperature and thermal preference (Jobling 1981). Pupfish (*Cyprinodon* spp.) have some of the highest lethal temperatures recorded, however data on their thermal preferenda are scarce. The objective of this study is to determine the final preferendum of the Tecopa pupfish, *C. nevadensis amargosae.*

Materials and methods

Fish were seined from the stream in March 1985 and June 1986. During the first occasion, we collected specimens from the outflow of Tecopa Bore and brought them to the University of Arizona (Tucson) where samples were acclimated to 25 and 35°C under the prevailing seasonal photoperiod prior to testing. Personnel of the National Park Service collected them the following year from the same location. They were kept at the University of Nevada in Las Vegas for a week and then shipped by air to the Waterloo laboratory in Ontario, Canada.

At Wilfrid Laurier University, a breeding colony was established from which individuals of the F_1 generation were randomly selected as experimental fish. They were maintained at 27°C, a temperature shown to be most favorable for reproduction (Gerking 1981). They were fed mixtures of commercial tropical fish food (tetramin), commercial trout food and frozen brine shrimp and maintained under a photoperiod of LD12:12. Specimens measuring 20–31 mm (total length) were tested in both a spatial and temporal thermal gradient.

Horizontal spatial gradient

Temperature preference of the 1985 sample was determined in the Tucson laboratory using a horizontal plastic trough (100 × 15 × 15 cm) in which temperature gradients were established by regulating heating and cooling at opposite ends through a thermostatically controlled immersion heater and a terminal compartment continually replenished with crushed ice. The trough was divided into nine equal cells by transverse transparent partitions containing apertures to allow free passage of the fish. Vertical thermal stratification was minimized by bubbling air into each of the cells.

The method in general consisted of tabulating the positions of fish in each of the nine cells at intervals of 30 seconds over a period of 10 minutes (designated a 'run'), after which the temperature gradient was reversed to eliminate orientation to nonthermal gradients and the procedure repeated. The average temperature of each cell during a run was calculated from initial and final temperatures as read from thermometers.

Lots of 10 fish were selected at random from the holding tank and transferred to the gradient trough. Experiments were conducted in the afternoon and early evening with 10 runs accomplished for the 25°C acclimated group and 8 for the group acclimated to 35°C. Frequency polygons describing the distributions of fish in chronological sequence of gradients were plotted, from which modal-selected temperatures were determined by inspection. Data from all runs were pooled and arithmetic mean occupied temperatures were calculated.

The data of runs for which the widest range of temperature was available were pooled and mean temperatures along with standard deviation were calculated.

Temporal gradient

Ten fish initially acclimated to 25°C were tested individually in an electronic shuttlebox which was a modification of that described previously (McCauley et al. 1977). Temperatures in this experimental aquarium were continuously monitored by a thermistor connected to a thermograph. Electronic shuttleboxes provide a continuous record of temperatures occupied by the animal throughout a 24 h period in a temporal gradient. This technique has been used notably by Reynolds & Casterlin (1979) as well as Beitinger (1981) to reveal diel rhythms in activity and temperature preference in fish.

Each fish was allowed 12 hours to habituate to the experimental situation before the heating and cooling systems were activated. Two additional days were then allowed for the fish to attain its final preferendum, a process shown to occur rapidly at high acclimation temperatures (Houston 1982).

Results and discussion

Behaviour in the spatial gradient

The experimental group of ten fish continuously swam up and down the length of the gradient trough encountering temperatures in the range 20–40°C. The 25°C acclimated fish were normally distributed at about 33°C (S.D. ± 2.5°C) and were

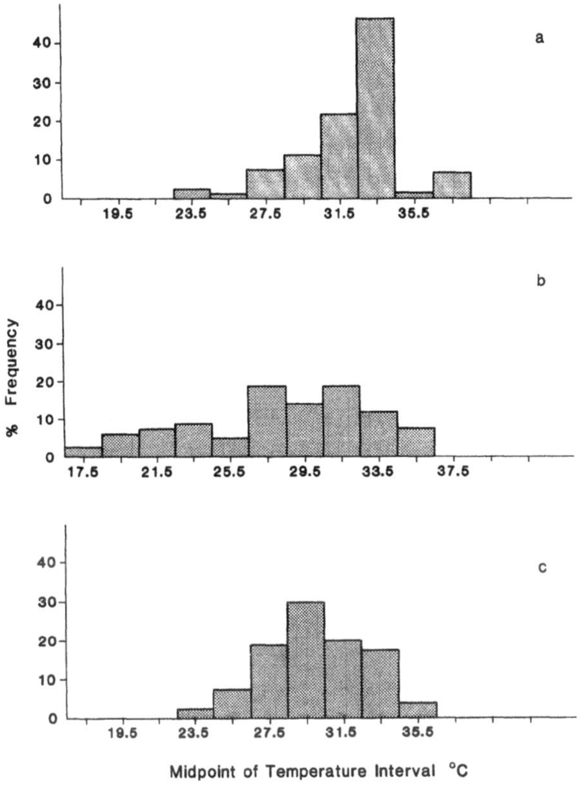

Fig. 1. Temperatures occupied by pupfish in laboratory thermal gradients: a– fish acclimated to 25°C and tested in a horizontal gradient, b– fish acclimated to 35°C and tested in a horizontal gradient, c– fish initially acclimated to 25°C and tested in an electronic shuttlebox over three days. Temperatures on the horizontal axes are the mid-points of 2°C intervals.

skewed towards the warm end of the trough. The group acclimated to 35°C, however, were less precise in selecting temperature and no distinct modal occupied temperature was apparent. Figure 1a and b summarizes the frequencies of temperatures occupied by fish acclimated to 25 and 35°C, respectively. Evidence that temperature was the dominant factor controlling the movements of the fish was achieved by changing the position of the preferred temperature in the trough and noting corresponding shifts in the modal fish position.

Behaviour in the electronic shuttlebox

Fish learned to control the temperature of the shuttlebox within 24 hours, whereupon they chose temperatures in the range of 25–35°C. The mean final preferendum of the ten individuals tested was 30.2°C (S.D. =/ 2.0°C).

Inspection of the time temperature curves of the thermograph tracings revealed no obvious diel rhythms in temperature preference.

Final preferendum

Mean acute preferred temperatures of fish acclimated to 25 and 35°C determined in the horizontal gradient trough were 31.5 and 28.2°C, respectively. That point at which acute preferred and acclimated temperature are coincident is the final preferendum (Fry 1947, Reynolds & Casterlin 1978). By interpolation of the curve relating acute preferred temperature to acclimation temperature, this value is 30°C which is in close agreement with that result obtained by the 'gravity method' i.e., allowing the fish to acclimate towards the final preferendum in the continuous thermal gradient provided by the shuttlebox. Both techniques indicate that the final preferendum for the species centers around 30°C.

As far as we know, the only other pupfish in which temperature preference has been studied is *Cyprinodon elegans*. Gehlbach et al. (1978) determined acute preferred temperature in a spatial thermal gradient similar to ours and found a preferred range of 21–25°C in the morning and 26–30°C in the afternoon. This progressive increase in temperature preference likely reflects upward acclimation of the fish in the gradient since experiments were carried out consecutively with the same group of fish (Gehlbach, personal communication).

Correlations of the final preferendum with other temperature dependent processes

There is a close correlation between the temperature at which maximum growth in a fish occurs and its final preferendum (McCauley & Casselman 1981, Jobling 1981, Houston 1981). However, Gerking & Lee (1983) found maximum growth temperature in *C. nevadensis nevadensis,* to occur at 22°C, some 8°C below the final preferendum of our fish. The optimum temperature range for reproduction, however, was 28–29°C.

Some validation of the final preferendum of 30°C is provided by the upper thermal tolerance for *C. n. nevadensis*. The upper incipient lethal temperature of this subspecies lies between 39°C and 40°C (Otto & Gerking 1972), a value characteristic of species having thermal preferenda at 30°C or slightly above (Jobling 1981, Houston 1981).

The thermal limits of heat tolerant fish

Those who have investigated the thermal tolerance of *Cyprinodon* acknowledge that this genus encompasses the most thermally resistant fishes that have been studied. Although the maximum final upper incipient lethal temperature appears to be 39.5°C, individuals inhabiting warm springs have been observed for short intervals at temperatures as high as 41°C. These sojourns must be of limited duration due to accumulating lethal thermal effects, but they do risk death to feed upon blue-green algae which flourish at these temperatures (Brown 1971). This risk-taking to feed at lethal temperatures is not restricted, however, to pupfish. Munson et al. (1980) demonstrated that hungry rainbow trout juveniles, would accept a 'thermal challenge' and successfully feed for short periods of time in the lethally warm end of an experimental trough when food was introduced.

Perusal of the literature reveals no fish species with upper incipient lethal temperatures above 40°C (Coutant 1972, Houston 1982, Jobling 1981). We propose that *Cyprinodon* is a member of a 'thermal guild' of species which have their upper thermal limit just below 40°C. This 'guild' of warm-water fish includes goldfish, carp, mosquitofish, and brown bullheads. The final preferenda of these species are clustered in the range 30–33°C, a value in agreement with the general rule proposed by Jobling (1981) relating final preferenda and upper lethal temperatures in fish.

We suggest that 40°C is likely the upper limit of the zone of the most thermally tolerant fish species, and that the optimum thermal habitat for growth, as reflected by final preferendum, lies some 10°C lower. Perhaps it is not a coincidence that 32°C is the mean July-August temperature of surface waters in the warmest part of the United States in summer (Miller et al. 1963). This generalization awaits the challenge presented by the discovery of a fish having a higher ultimate upper lethal temperature than that of the pupfish and other members of its thermal guild.

Acknowledgements

We thank J.E. Deacon (University of Nevada in Las Vegas) for encouragement and arranging to procure the specimens. Wayne Westphal of the U.S. National Parks Service collected fish and provided field observations. The project was supported by a generous summer research fellowship from Wilfrid Laurier University. Summer students Bill Barg and Cathy Tash assisted in experiments at the University of Arizona.

References cited

Beitinger, T.L. 1981. Diel activity rhythms and thermoregulatory behavior of bluegill in response to unnatural photoperiods. Biol. Bull. 149: 96–108.

Brown, J.H. 1971. The desert pupfish. Sci. Amer. 225: 104–110.

Coutant, C.C. 1972. Time-temperature relationships for thermal resistances of aquatic organisms, principally fish. Oak Ridge National Laboratory (ORNLEIS) 72–27, Oak Ridge. 100 pp.

Deacon, J.E. & W.L. Minckley. 1974. Desert fishes. pp. 385–487. *In:* G.W. Brown Jr. (ed.) Desert Biology, Vol. 2, Academic Press, New York.

Fry, F.E.J. 1947. Effects of the environment on animal activity. Univ. Toronto Stud., Biol. Ser. 55, Publ. Ont. Fish. Res. Lab. 68: 1–62.

Gehlbach, F.R., C.L. Bryan & H.A. Reno. 1978. Thermal ecological features of *Cyprinodon elegans* and *Gambusia nobilis,* endangered Texas species. Texas J. Sci. 30: 99–101.

Gerking, S.D. & R.M. Lee. 1983. Thermal limits for growth and reproduction in the desert pupfish *Cyprinodon n. nevadensis,* Physiol. Zool. 56: 1–9.

Gerking, S.D. 1981. The contribution of the desert pupfish (*Cyprinodon nevadensis*) to fish reproduction in stressful environments. pp. 335–356. *In:* R.J. Naiman & D.L. Soltz (ed.) Fishes in North American Deserts, John Wiley & Sons, New York.

Houston, A.H. 1982. Thermal effects on fishes. Publication 18566, Environmental Secretariat, National Research Council of Canada, Ottawa. 200 pp.

Jobling, M. 1981. Temperature tolerance and the final preferendum – rapid methods for the assessment of optimum growth temperatures. J. Fish Biol. 19: 439–455.

McCauley, R.W., J.R. Elliott & L.A.A. Read. 1977. Behavioral thermoregulation by rainbow trout (*Salmo gairdneri*) at various acclimation temperatures. Trans. Amer. Fish. Soc. 106: 362–365.

McCauley, R.W. & J.M. Casselman. 1981. The final preferendum as an index of the temperature for optimum growth in fish. pp. 81–93. *In:* K. Tiews (ed.) World Symposium on Aquaculture in Heated Effluents and Recirculation Systems, Stavanger 28–30, May 1980, Volume 2, Heinemann, Berlin.

Miller, R.R. 1948. The cyprinodont fishes of the Death Valley of eastern California and southwestern Nevada. Misc. Publ. Mus. Zool. Univ. Michigan, Ann Arbor 68: 1–155.

Miller, D., J.J. Geraghty & R.S. Collins. 1963. Water atlas of the United States. Water Information Center, Port Washington. 40 pp.

Munson, B.H., J.H. McCormick & H.L. Collin. 1980. Influence of thermal challenge on conditioned feeding of juvenile rainbow trout. Trans. Amer. Fish Soc. 109: 116–121.

Naiman, R.J. 1976. Primary production, standing stock and export of organic matter in a Mohave desert thermal stream. Limnol. Oceanogr. 21: 60–73.

Otto, R.G. & S.D. Gerking. 1973. Heat tolerance of a Death Valley pupfish (Genus *Cyprinodon*). Physiol. Zool. 46: 43–49.

Reynolds, W.W. & M.E. Casterlin. 1979. Behavioral thermal regulation and the 'final preferendum paradigm'. Amer. Zool. 19: 211–224.

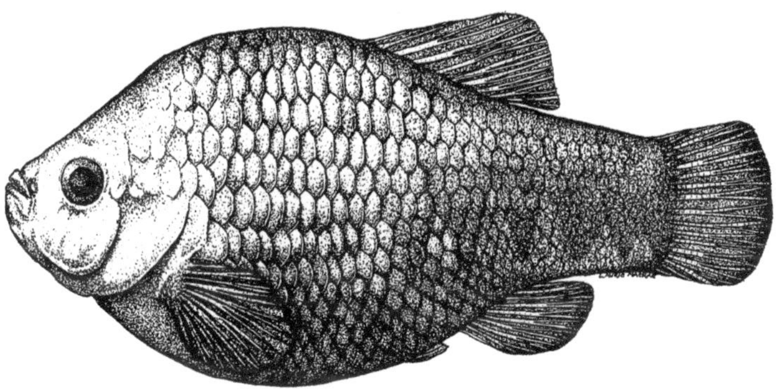

Environmental Biology of Fishes Vol. 23, No. 1–2, pp. 141–154, 1988

The need for conservation and management of Philippine coral reefs*

Peter J. Rubec
International Marinelife Alliance Canada, 355 Route de la Mer, B.P. 4, Sainte-Flavie, Quebec GOJ 2LO, Canada

Received 21.8.1987 Accepted 18.1.1988

Key words: *Cyanide, Diversity, Development, Ecology, Environment, Fisheries, Pollution*

Synopsis

The biological diversity and productivity of Philippine coral reefs are threatened by siltation, destructive fishing methods, coral and shell collecting and overfishing. Destructive fishing includes: the widespread, illegal use of explosives; poisons such as sodium cyanide; muro-ami and kayakas fishing; and trawling. The recent decline in catch rates threatens the livelihood of 700 000 near-shore subsistence fishermen who catch 55% of the total landings. The new government under Corazon Aquino wishes to protect the marine environment through the creation of a viable marine conservation management plan, enforcement of existing laws and through cooperation between government, non-government and international agencies to provide education and research. The Department of Agriculture, which has the authority for fisheries, has designated the International Marinelife Alliance as the lead non-government agency involved with fund raising, net-training and finding alternatives to destructive fishing methods.

Species diversity

The world is undergoing the largest extermination of animal and plant species in its history (Tangley 1985). The destruction of tropical rain forests and their associated terrestrial fauna and flora have been widely publicized. Less attention has been paid to freshwater and marine environments in the tropics, which are also threatened by human activity. This paper reviews the factors causing the destruction of Philippine coral reefs and their associated biota.

Since World War II, as human populations have grown, human activities have placed increasing pressure on coral reefs world-wide. Various forms of reef degradation were described by the participants at the Fourth International Coral Reef Symposium held in Manila (Stoddart 1981). Philippine coral reefs were described as being the most abused in the world (Gomez et al. 1981).

Coral reefs within the 40 fathom (= 219 m) contour cover an estimated 44 000 km^2 interspersed among 7 100 islands of the Philippine archipelago (Bina 1982). Considering that most corals occur within the 20 fathom (= 110 m) depth contour, the area of the reefs would be 33 000 km^2 (Carpenter 1977). While the exact area of coral reefs is not known, the area with hard substrates within the 20 fanthom contour may be close to 27 000 km^2 (Murdy & Ferraris 1980, Silvestre & Ganaden 1987).

Both terrestrial and marine habitats (Fig. 1) in the Philippines have high biological diversities,

* This essay addresses the societal relationships of village fishery to government, aid organizations, and science. Vadim Vladykov always maintained close relationship with fishermen and devoted over a year of his life to helping develop fisheries in Iran.

Fig. 1. Underwater reefs are productive oases in otherwise desert-like tropical waters. Philippine reefs have a higher diversity of fishes than the Australian Great Barrier Reef. Photo by L. Funkhouser.

which are suffering from a lack of environmental protection. A total of 488 species of coral in 78 genera have been described (Nemenzo 1981). A survey of benthic algae found 971 species in 209 genera (Silva et al. 1987). Thirteen species of seagrasses have been described (Menez et al. 1983). There are over 1700 mollusc species (Springsteen & Leobrera 1986). The Philippine islands have the highest marine fish species diversity (more than 2000) of any reef areas (Herre 1953, Sale 1980). A *Philippine National Conservation Strategy* (PNCS) was proposed in 1983 to protect biological diversity and manage ecosystems; but most of the recommendations have not been implemented.

Productivity

Nybakken (1982) noted that coral reefs are oases in a watery desert. Tropical marine waters tend to be extremely poor in nutrients, and hence low in productivity. Yet, coral reefs abound with life and all studies have indicated very high primary productivity. The secret of the reefs is their apparent ability to hold nutrients in the system and to act as a sink for those brought in from outside. This appears to be largely due to the ability of photosynthetic zooxanthellae. Corals and giant clams lead a largely autotropic existence by obtaining nutrients from symbiotic zooxanthellae (Goreau et al. 1979). In addition, benthic algal turfs sustain high primary production on coral reefs (Adey 1987). The zooxanthellae and benthic algae support a food chain on which man is heavily dependent.

Inshore areas including coral reefs, seagrass beds and mangroves account for 60% of the total fish catches, as well as fishing products derived from the reefs such as dried sea cucumber, stony and precious corals, invertebrates, seaweeds and seashells (Gomez et al. 1981, Bina 1982). Among the islanders of the southern Philippines, coral reefs are the source of 80% of their diet. The Philippines has historically relied on fish for the bulk of its animal protein. The per capita fish consumption in the Philippines was 41 kg yr^{-1} in 1984 (BFAR 1984). High fish catches have been reported from recent surveys for the Island of Sumilon (36.9 mt km^{-2} yr^{-1}) and Apo Island (31.8 mt km^{-2} yr^{-1}) respectively off the islands of Cebu and Negros (Acala, unpublished, White & Savina 1987). These islands represent relatively pristine coral reef environments which have benefitted from protective management. Elsewhere in the Philippines, coral habitats and mangroves important for fish production are greatly affected by human activities. Fish catches from two degraded reefs (Selinog Island in the Mindanao Sea and Hulao-hulao in the Panay Gulf, respectively) were 5.9 and 5.0 mt km^{-2} yr^{-1} (Luchavez et al. 1984). These reefs have been dynamited and subjected to muro-ami fishing.

Reef degradation

The degradation of coral reefs due to human activities is common throughout Southeast Asia, but most acute in the Philippines (Gomez 1980). A nation-wide survey of Philippine coral reefs found that 70.6% of the reefs were in poor to fair condition (Gomez et al. 1981). Out of 559 stations, the results were 31.8% poor, 38.8% fair, 23.7% good and 5.7% excellent. The causes cited were widespread siltation and sedimentation due to forest denudation and the prevalence of destructive fishing methods.

A special committee prepared a comprehensive review on the condition of coastal fishery resources in 1981. A revised report *The Central Visayas Resource Management Project* (CVRMP) was issued in 1984. This report noted that continual assaults by human activities have put the fisheries under severe strain. It was noted that many reefs were barely recruiting fishes. From 1976–1980 commercial food fish had declined by 25% in Central Visayas Region VII. It noted a declining catch in five groups that together account for over 90% of the commercial harvest. Group 33, which had a 35% decline in catch during this period, is composed of many desirable species dependent on mangrove and coral reefs. The situation appears to be similar in other parts of the Philippines (Smith et al. 1980). Recent reports indicate that there is overexploitation of near-shore fishery resources (Abrea et al. 1986, Mines et al. 1986, Silvestre & Ganaden 1987). The trend towards overfishing was noted to be a matter for immediate concern (Smith et al. 1980). Silvestre et al. (1986) estimated that the biomass of near-shore (0–100 m depth) demersal stocks have declined to about 30% of their biomass in the late 1940's. The economic loss that such lack of management entails could be as much as US $90 million per year (Silvestre & Ganaden 1987).

Deforestation has contributed to coral reef degradation. Illegal lumbering has long been a problem and starting in 1973, President Marcos granted timber concessions to large corporations including Cellophil (Bello et al. 1982, Zich & McCurry 1986). Officials of the Bureau of Forest Development say that 11.1 M ha of Philippine forest land are still covered with timber (Cruz 1986). However, satellite data and surveys conducted by other agencies reveal a smaller figure. According to the University of the Philippines Forestry Development Center, the remaining virgin forests in the Philippines in 1986 was between 1.9 and 2.4 M ha. In the 1970's, the Bureau of Forest Development estimated that the rate of forest destruction was 170 000 ha yr^{-1} (Alvarez 1984). However, a 1977 study by the National Environmental Protection Council revealed that the Philippines had been losing its forests at the rate of 700 000 ha yr^{-1} since 1974.

Siltation caused by logging and other activities is regarded by marine biologists are the primary cause of coral reef destruction in the Philippines (Gomez et al. 1981). By cutting light penetration and smothering coral, the productivity of the system is affected (E. Gomez, personal communication 1986). An attempt to correlate the degree of forest denudation with turbidity of adjacent coastal waters was mentioned by Bina (1982), but I am not aware of any scientific reports which quantify the extent of the effects of siltation in the Philippines.

Destructive fishing methods

Destructive fishing practices plaguing the Philippines include fishing with explosives or 'blast fishing', the use of sodium cyanide, kayakas and muro-

Fig. 2. Over 70% of Philippine coral reefs are in poor to fair condition due to sedimentation and destructive fishing methods. This shows a reef destroyed through explosives and cyanide. Photo by L. Funkhouser.

ami, trawl fishing and purse seining (Cruz 1986, Rubec 1986). These practices are often conducted over the same areas, making it difficult to separate and quantify their relative degrees of damage to the coral reefs. It is probable that the effects are synergistic. Many laws exist against destructive fishing methods. But, it is only under the recent Aquino government that the laws have started to be enforced.

Use of explosives. Blast fishing to gather food fish using bottles containing dynamite or ammonium nitrate is widespread in the Philippines (Alcala & Gomez 1979, Randall 1983). Up to 50% of the 700 000 small-scale fishermen may use explosives to catch fish for at least part of the year. According to recent newspaper reports, the Philippine military is one of the groups involved in the illegal sale of explosives to fishermen. Some military personnel may also be involved in the purchase, transport and marketing of fish caught by blast fishing. Fishing with explosives was banned in 1972 (Presidential Decrees 1081, 1058). But, it continues unabated in many areas.

The use of explosives kills fish non-selectively, including young fish that are too small to be sold or eaten (Cruz 1986). Newspaper articles in the Philippines reported that an estimated 5–10 mt of blast captured fish are sold in the Manila market each day. The long term use of explosives has reduced many coral reefs to rubble in the Philippines. Most explosives are used in shallow water or near the surface (S. Robinson, personal communication 1985). During 1986, I photographed dead, shattered, grey-coloured coral heads off Bolinao, Province of Pangasinan.

The Danajon Bank, in the Central Visayas between the islands of Cebu and Bohol is one of the largest coral reef areas in the Philippines (1715 km^2), which was noted for its high fish harvest (CVRMP 1984). The decline of the harvest on the Danajon Bank began in the 1950's with extensive dynamite usage (McLarney 1986). From an underwater perspective, it is a disaster area. In many places, hectares of coral have been toppled and shattered by explosions (Fig. 2). Most of these prostrate corals, the standing fragments and the

Fig. 3. Aquarium fish collector with plastic squirt bottle containing cyanide solution which is sprayed on the coral head to stun the fish. Photo by L. Funkhouser.

remnants of structurally intact reef are dead and covered with algae. Coral fragments wreck additional havoc when moved by storm waves. Coral planulae, which need firm substrates to settle, cannot colonize coral rubble (A. Alcala, personal communication 1987). A mean recovery time of 38 years was estimated for 50% regeneration of dynamited coral reefs (Alcala & Gomez 1979). Dynamite does structural damage, but leaves enough living coral to repopulate the reef (McLarney 1986). Food fish desert a dynamited reef, but smaller aquarium fishes persist in the area.

Cyanide fishing. Sodium cyanide (NaCN) was introduced for the capture of marine fish for the ornamental fish hobby in 1962. The use of cyanide by both fishermen and tropical fish collectors has transformed vast tracts of once-healthy, living reefs into dead, decaying waste-lands (Cruz 1986). Cyanide is a very broad spectrum poison acting against the enzyme systems responsible for respiratory metabolism (Rubec & Pratt 1984, Rubec 1986). Despite being illegal (Presidential Decrees 704, 1058), cyanide collecting is still widespread.

Fish collectors use plastic bottles containing dissolved NaCN in the form of hydrocyanic acid (HCN) to kill food fish and to stun aquarium fishes (Rubec 1986). As an aquarium fish collector approaches, the fish seek refuge in the coral. The diver seals off all the exits by squirting clouds of HCN onto the coral head (Fig. 3). Not being able to control the concentration from the squirt bottle, about 50% of the exposed fish die entombed in the coral. The remaining fish flee; but many are stunned by the HCN. The aquarium fish collector selects about 10% of the exposed fish, taking the colourful species of interest to aquarists. The remaining fish lying on the bottom may be collected for the dinner table or left for scavenging predators.

Food fish captured using NaCN are often sold in the marketplace (Cruz 1986). A clandestine trade exists in the capture of live food fish using NaCN, which are smuggled to Hong Kong (S. Robinson, personal communication 1986). The main species collected are the polkadot grouper, *Cromileptes altivalis,* and the giant labrid *Cheilinus undulatus.*

Larger ships with as many as 50 divers, use up to 1250 kg of NaCN for a single 10–20 day fishing trip off the Island of Palawan. The fishery has operated for at least 15 years.

The fish which survive the initial cyanide exposure, exhibit delayed mortalities. More than 80% of the aquarium fishes captured die throughout the chain from the collector to the marine hobbyist (Rubec 1986). Cyanide exposure has been shown to cause internal damage to the fishes' liver, intestine and reproductive organs. The exposed fish which escape on the reef are probably similarly affected; having impaired survival and reproduction.

Data obtained from the Coral Reef Research Section of the Philippine Bureau of Fisheries and Aquatic Resources (BFAR) indicates that the repeated use of NaCN kills coral heads (Rubec 1986). This was confirmed by the aquarium fish collectors interviewed. Initially, the coral heads take on a bleached-white appearance before they become coated with algae.

Robinson believes that it was the use of cyanide, patiently and repeatedly applied to one coral head after another, which erased not only aquarium fishes, but virtually the entire support system from small to large fishes (Robinson 1984, McLarney 1986). There are about 1500 aquarium fish collectors and possibly another 1500 food fishermen using NaCN (V. Albaladejo, personal communication 1986). An active collector douses 50 coral heads a day for about 225 days out of the year. It is estimated that over 33 million coral heads are sprayed with cyanide in the Philippines each year. Cyanide destroys the natural habitat on which reef fish depend for food, shelter and reproduction. Its use in the aquarium and food fish trades, threatens the reef ecosystem.

Muro-ami and kayakas fishing. Another destructive fishing method, muro-ami was introduced by the Japanese in the 1920's (Montalban 1927). Annual catches by muro-ami from 1960–1975 ranged from 9 362–26 475 mt (Carpenter 1977). It consists of setting a net 6–25 m deep near the reefs (Hall 1985). A line of about 300 boys age 7 to 15 years is used to drive the fish into the net. Each boy is equipped with a scare-line consisting of a length of rope tied with plastic strips and a stone weight. When the swimmer jiggles the scare-line, the stone bounces off the coral. The resulting sound vibrations and the wall of plastic strips drive the fish out of the coral towards the net.

Two companies involved with muro-ami are the Abines Muro-ani Fishing Corporation and Frabel. The latter operates more than 30 vessels, each of which may carry a crew of about 350 men and boys. Each boat nets about 10 000 kg fish per day. The poorer quality fish are eaten by the crew, medium quality is salted on board and the high value fish are packed on ice and sold at the nearest marketplace.

The catch consists of a variety of schooling herbivores and planktivores (Caesionidae, Acanthuridae, Carangidae, Lutjanidae, Lethrinidae), since most bottom associated and presumably territorial fishes ignore the scare-lines (Carpenter & Alcala 1977). From 1960–1975, in all except three years, muro-ami boats had the highest annual catch (179–804 mt yr^{-1}) per unit vessel (Carpenter 1977).

While one pass over a coral reef may not appear to cause much coral reef damage (Hall 1985), this does not consider that muro-ami operations are repeated in the same areas year after year. Studies have shown that considerable coral is lost (Carpenter & Alcala 1977). A report by the Special Committee for the Development and Promotion of Underwater Diving and Tourism condemned muro-ami because it destroys living colonies of coral polyps. About 36 000 ha of coral reef are damaged each year (M.T. Logarta, *The Manila Chronicle,* July 1, 1986). Philippine newspapers during 1986, denounced the destruction of the reefs and the exploitation of children by muro-ami vessels. In early December 1986, muro-ami and kayakas fishing were banned by Fisheries Administrative Order 163.

Kayakas fishing is similar to muro-ami, but occurs in shallower water (Carpenter & Alcala 1977). Swimmers form a line and drive fish towards a net by smashing the corals with poles. The breaking of corals during kayakas seems unnecessary. The noise and scare-line were observed to be the major factors in driving the fish towards the net. Catch composition varies widely with Scaridae, Labridae,

Acanthuridae and Syganidae forming the bulk of the catch. This gear harvests an average of 1.8 mt d^{-1}. Mobile vessels with kayakas divers raid reefs, leaving a trail of destruction for local residents. Carpenter & Alcala (1977) concluded that both muro-ami and kayakas caused considerable damage to the corals. The breaking of corals was considered detrimental to long term fisheries production and unnecessary to these gears' efficiency.

Trawling and seining. Trawling and purse seining are fishing methods which also cause the destruction of coral reefs. There are laws in the Philippines controlling the number of vessels by licenses and restricting the areas fished. During 1986, charges were laid against fishermen and their boats seized for illegal fishing with trawls. In addition to violations by domestic vessels, it was revealed that 17 ultra-modern trawlers owned by Taiwanese businessmen had been operating for the last four years in Philippine waters without commercial fishing licenses (M. Anulan, *Tempo,* Aug. 2, 1986). The Philippine Deep-Sea Fishing Association has been exposing the massive entry of Taiwanese and Japanese poachers into Philippine waters, thus posing stiff competition to Philippine fishermen and depriving the small coastal fishermen of their usual catch. Competition between near-shore trawlers and small-scale fishermen has been studied in San Miguel Bay by Mines et al. (1986).

A large variety of fishing methods are in use in the Philippines (Smith et al. 1980). It is a classic case of Hardin's Tragedy of the Commons. Illegal fishing practices, such as dynamiting and the use of poison, are common as effort intensifies for limited and declining resources. Habitat destruction is extensive and it may have influenced demersal fisheries catches more than overfishing.

Mangrove nursery areas

Research results link the mangrove ecosystem with the coral reef ecosystem (Pangilinan 1984). Many reef fish use mangroves as nursery areas. Mangroves provide habitat for crustaceans and molluscs. They stabilize shorelines and provide fuel and building materials to local people. It is probable that close to 1 M ha of mangrove forests surrounded most Philippine islands in the late 1800's (Alcala, personal communication 1987). In a nationwide survey conducted in 1980, using satellite imagery and aerial reconnaissance, it was found that only 146 000 ha remained (Bina & Roque 1980). In 1984, there was only a little over 100 000 ha left (Pangilinan 1984). The rest was logged for export, transformed into milkfish ponds and salt beds or reclaimed for housing projects. This prompted the government to issue two presidential decrees (704, 705) and three conservation decrees (950, 1121, 1152) to conserve and preserve mangrove forests. A total of 70 000 ha were alloted for conservation and 4 500 ha, mostly consisting of mangrove islets, were declared as nature preserves (Bina 1982). Reforestation programs by the Bureau of Forest Development and the Central Visayas Management Project have shown that it is possible to restore mangroves at low cost (Yao 1987). Despite this, mangroves are still being destroyed at the rate of 17 000 ha yr^{-1}, according to statistics of the Bureau of Forest Development (Panaranda 1984).

Ornamental coral and shell collecting

During the 1970's the Philippines was the source of more than 90% of the ornamental coral imported by other countries worldwide (Wells 1981). Exports increased to a peak of over 2 000 mt in 1976. Presidential Decrees in 1977 and 1980 (1219, 1698) banned the export of unworked stony corals (Radan 1984, Alvarez & Fleras 1985). The ban in 1977 caused coral exports to drop to 170 mt, but lax enforcement allowed the estimated exports to rise to 962 mt in 1978 (Wells 1981). Despite the laws, coral gathering for export is still common in the Philippines (McManus 1980, Alvarez & Fleras 1985).

Coral gathering is a major source of income for 4 000 families or 30 000 people in Bohol, Zamboanga City, Tawi-Tawi, Surigao and Palawan (Alvarez & Fleras 1985). Illegal gatherers harvest about 600 mt yr^{-1} of ahermatypic (precious) corals

Fig. 4. Boxes of corals at a Florida curio shop labelled as coming from the Philippines, despite the fact that such exports have been illegal since 1977. Photo by L. Funkhouser.

and about 48 000 mt yr^{-1} of hermatypic (stony) corals. Large quantities of corals are being smuggled out of the Philippines from the islands of Mindanao and Cebu to Australia and New Caledonia for shipment elsewhere. Four large container vans of corals are shipped to Europe monthly (Fleras 1984).

The 48 000 mt yr^{-1} estimate of corals gathered in the Philippines (Alvarez & Fleras 1985) initially sounds like a gross overestimate. It exceeds by a factor of 50 the estimate of coral exports for 1978 reported by Wells (1981). While some of this is used domestically for construction of roads and buildings, it seems very likely that the illegal export trade in coral far exceeds what has been estimated (Wells 1981). Wells (1982) visited the Philippines to obtain information on the scale of the trade. She found that corals were exported labelled as shells or shellcraft or as fillers in consignments of other items (Fig. 4). The U.S.-Philippines coral trade appears as active today as at any time in the last few years (Fuller 1986).

Wells (1981) noted that the ornamental shell trade is closely linked with the stony coral trade (Fig. 5). The Philippines is currently the main supplier of ornamental shells world-wide. It also exports large quantities of shellcraft items, many of which include mother-of-pearl and the windowpane oyster, *Placuna placenta.* Exports increased during the 1970's and peaked in 1978 when 2000 mt of shellcraft items and at least another 100 000 shell articles were exported. Imports into the United States of articles made of shells peaked in 1976, with about 75% coming from the Philippines. Since then, shell imports from the Philippines have declined, possibly due to a decline in the abundance of molluscs.

It is apparent that coral gathering causes habitat destruction that takes many years to regenerate. There is no evidence that the commercial collection of corals or marine molluscs have yet caused any of them to become extinct (Wells 1981). However, the current heavy demand and the numerous reports of depleted populations give cause for con-

Fig. 5. A shell collecting boat used to gather mollusc shells for export from the Philippines. Photo by L. Funkhouser.

cern. There is a pressing need for studies of the ecology of marine invertebrates in the Philippines, in order to ascertain sustainable yields for food and for the ornamental coral and shell trades.

Conservation measures for coral reefs

The preceeding text has described the importance of coral reefs in sustaining marine productivity on which Filippinos are heavily dependent. There is evidence that reefs destroyed by dynamite regenerate very slowly (Alcala & Gomez 1979). Measurements of coral growth rates for various species range from a few millimeters to 18 cm yr^{-1} on a linear axis (Gomez et al. 1985). The degradation of Philippine coral reefs has resulted in declining catch rates for fishes. Conservation measures are urgently needed.

Unlike other Pacific Islands, the Philippines has no conservation tradition of marine resources tied to local customs and religion (Salvat 1981). Since World War II a shift from a subsistence economy to a western consumer-oriented economy has occurred (Randall 1983). The rising human population (57 million) and the open access nature of the resource has encouraged surplus fishermen; which in turn has increased fishing effort and hastened the depletion of marine resources. Unfortunately, it is not the village fishermen who have benefited from overexploitation of marine resources (Robinson 1987).

While it is impossible to return to the original subsistence economy, many of these concepts should be enacted in a viable Philippine coral reef management plan. No plan presently exists. Coral reef parks can play an important role in preserving biological diversity, increasing fish yields and providing jobs and income from tourism (Salvat 1981, White 1986a, b). Marine parks have been created in the Philippines (Castaneda & Miclat 1981, Cabanban & White 1981). There are at least 19 marine parks, or reserves but most are neither developed nor properly managed due to sociological, eco-

Fig. 6. The International Marinelife Alliance has demonstrated that collecting marine aquarium fish with fine-mesh nets is a viable alternative to the use of sodium cyanide and it is seeking funds for a nationwide net-training program. Photo by L. Funkhouser.

nomic and political problems. Carbin Reef (off Negros), and the Islands of Sumilon (off Cebu), Apo (off Negros), Balicasag (off Bohol) and Pamilacan (off Bohol) have received some protection through local community involvement and municipal ordinances (Alcala unpublished, White & Savina 1987).

The Philippine government of Corazon Aquino has recognized that marine conservation is an important priority. Ramon Mitra, the minister of Agriculture and Food (DA) in charge of fisheries issued the following statement in 1987:

> 'The rapid depletion of our fishery resources due to destructive fishing practices, particularly the use of explosives and poison, is threatening to aggrevate problems of poverty in rural fishing areas. The magnitude of the problem and its consequences of malnutrition and hunger has prompted me to create a Task Force on Illegal Fishing with sufficient power and authority to arrest and prosecute violators of fishery laws . . .
>
> The Task Force under the able leadership of the Philippine Fisheries Development Authority (PFDA) has proven itself equal to the task. But insufficiency of funds and logistics, on top of existing loopholes in the legal system, is a major handicap that stymies further efforts at greater accomplishments.'

The Task Force on Illegal Fishing headed by Malcolm Sarmiento of the PFDA and assisted by the Coast Guard and BFAR has been enforcing the laws against fishing with explosives and cyanide. Philippine newspapers carried numerous stories on arrests which were made during the summer and fall of 1986.

There is a DA Coordinating Committee on Illegal Fishing which consists of government and non-government agencies which are working to coordinate and publicize problems associated with destructive fishing and seeking alternative methods. On the committee are representatives of the DA, BFAR, PFDA, Agricultural Training Institute, International Marinelife Alliance (IMA), Federation of Fishing Associations of the Philippines, Philippines Tropical Fish Exporters Association and RBS and Co.

The International Marinelife Alliance is a non-profit organization concerned with conservation of the diversity of sea life, protection of marine environments and promotion of sustainable use of marine resources. The IMA has been effective in harnessing public awareness on the extent of illegal fishing in the Philippines, directing and coordinating a pilot program to train divers in conservation principles and the use of hand nets and barrier nets (Fig. 5), and by providing international funding support for the net-training program, intended to provide a viable alternative to cyanide use for aquarium fish collection. In December 1986, the IMA was designated the lead non-government agency involved with fund raising, and to direct and coordinate the nationwide net-training program for cyanide users in cooperation with concerned agencies in the DA.

The IMA has proposed a three-year marine environmental program for the Philippines involving activities in coral reef management, alternatives to dynamite and muro-ami fishing, environmental assessment, baseline taxonomic surveys, fishermen's cooperatives and publication of anti-destructive fishing information materials. Ramon Mitra on behalf of the DA has endorsed the proposal and considers this initiative as a major input to the government's comprehensive anti-destructive fishing program.

The need for a village-based management plan

Fishery management policies have been geared to the large fishing companies which exploit offshore species such as tuna for export. Little attention has been given to small-scale municipal fishermen in coastal villages, who comprise 90% of all fishermen and produce about 55% of the total catch (Smith et al. 1980). These people are the poorest socio-economic group in the Philippines. Net annual mean income for a subsistence fisherman was 2405 Philippine pesos (which is now about $120 U.S.) (Librero et al. 1985). Since 1945, because of destructive fishing methods and competition from larger boats fishing near-shore, the municipal fisherman's daily catch has dropped from 7–10 kg d^{-1} to 3–5 kg d^{-1} (Carolino 1987). The small-scale fisherman is presently catching barely enough to feed his family. Many have very little extra fish to sell in order to obtain cash income. There is a need for an overall resource management plan which protects marine habitat and resources near-shore and addresses the needs of village fishermen (Emmerson 1980).

An infrastructure must be built which allows collaboration between Philippine government agencies, universities, international agencies, conservation organizations and especially the various user groups including fishermen, fish collectors, coral gatherers, aquaculturists, etc. The Coordinating Committee on Illegal Fishing is a positive step in this direction.

The establishment of a workable marine resource management plan will require extensive consultation with the various agencies and user groups. The plan should include marine education programs, multiple use zoning, regulatory measures and mechanisms to ensure enforcement, marine research in fisheries and aquaculture, extension programs to transfer knowledge and techniques to and between user groups and mechanisms built into the system for communication between all parties on a continuing basis. The successful implementation of this strategy hinges largely on the inclusion of conservation principles as a major factor in the planning of all sectors. It is hoped that the Marine Resource Management Plan will be part of a larger Philippines National Conservation Strategy (PNCS 1983).

The absence of coordination in the implementation of conservation projects in the Philippines is largely responsible for the previous lack of success in conservation efforts and the failure of conservation principles to be integrated into development (PNCS 1983). It is necessary that organizations representing the local people should have a role in defining the problems and seeking solutions. It is only through their support and cooperation that village based conservation and development programs will succeed. Better results may accrue if they are given responsibilities for local resource management schemes (Bojos et al. 1987, Robinson 1987).

Since coastal fisheries are overexploited, there is a need to develop alternative means for coastal fishermen to make a living. Mariculture of fish, molluscs, crustaceans and seaweeds can expand (Doty 1983), provided these do not destroy the habitat of adult or juvenile fish and invertebrates. The extensive culture of giant clams is now feasible (Heslinga & Fitt 1987). Tourism including recreational fishing and diving can be developed in certain areas in combination with nature reserves and parks, which may help in reestablishing the marine fauna. The making of handicrafts, which don't use marine organisms, could also be implemented. The marine theme could still be emphasized through paintings, prints, ceramics, weaving, etc. The productivity of the coral reefs should increase, if effort from human exploitation is reduced.

Local cooperatives for fishers, aquarium fish collectors, handicraft makers, aquaculturists etc., would give the local people the control of marketing their products. Through local collective concern for the environment and for products from the marine environment, there should be a higher economic return. Buyers of marine fish for food and aquarium fish for export should be willing to pay more for better quality products. Reduced mortalities of net-caught aquarium fish will mean more profitability for exporters in Manila. A national fishermen's newspaper would serve to inform fishermen of markets and prices paid in different parts of the country, and about better techniques and gear.

The destruction of Philippine coral reefs described in this paper has sociological ramifications. Marine destruction translates into malnutrition of local people and the nation which is fed by fishermen. The Philippine government is very concerned, but it will need funding and technical assistance from other countries to deal with the problems. Coral reefs are a heritage which belong to the world. The solutions found in the Philippines can be used elsewhere, where similar problems already exist with coral reefs.

The species diversity of Philippine coral reefs is still not completely known. The collections of fishes (Herre 1953) and other organisms which existed in the Philippines were largely destroyed during the Second World War. There is a need to establish systematic collections of species as part of biological baseline studies (Wilson 1985). In the Philippines, protecting biological diversity is of immediate economic and social concern. The coral reefs support the village fishermen, their families and a large part of Philippine society.

Acknowledgements

S. Robinson provided his first hand knowledge obtained from living with village fishers in the Philippines. I had the privilege to interview V. Carolino, a fisherman in Binabalian on Santiago Island. He acted as translator for interviews with four divers, J. Genoso Jr., A. Purgatorio, G. Gornie, N. Caasi, who formerly used cyanide to collect fish and now use reef nets. D.E. McAllister, V.R. Pratt and S. Robinson accompanied the author to the Philippines where they met with officials in the new government under President Corazon Aquino in June 1986. Minister R. Mitra of the DA and his staff were most helpful in explaining concerns about the coral reefs and the need to help village fishermen. E.D. Gomez of the University of the Philippines and A. Alcala of Silliman University provided valuable research studies. I.A. Ronquillo, V. Albaladejo and R. Miclat of BFAR provided useful data. L.H. Matienzo Jr., of Greenfields magazine provided valuable information reflected in the references.

References

Abrea, R.A., N.C. Mendoza, C.S. Reyes, C.A. Angsinco, R.E. Ancuna & M.M. Tan. 1986. Small scale fisheries in Iligan Bay, Philippines. pp. 341–348. *In:* J.L. Maclean, L.B. Dizon & L.V. Hosillos (ed.) The First Asian Fisheries Forum, Asian Fisheries Society, Manila.

Adey, W.H. 1987. Food production in low-nutrient seas. BioScience 37: 340–348.

Alcala, A.C. & E.D. Gomez. 1979. Recolinization and growth of hermatypic corals in dynamite-blasted coral reefs in the Central Visayas, Philippines. Proc. Int. Symp. Mar. Biogeogr. Evol. S. Hem. 2: 645–661.

Alvarez, R. 1984. Our vanishing forests. Greenfields 14(2): 6–16.

Alvarez, R.C. & J. Fleras. 1985. The wanton destruction of Philippine coral reefs. Greenfields 15(1): 36–40.

Bello, W., D. Kinley & E. Elinson. 1982. Development debacle: the world bank in the Philippines. Institute for Food and Development Policy, San Francisco. 264 pp.

Bina, R.T. 1982. Application of LANDSAT data to coral reef management in the Philippines. Remote sensing of the Great Barrier Reef Region, Workshop Proc., Townesville, Australia, May 5–7, 1982: 1–38.

BFAR 1984. Fishery Statistics 1984 of the Philippines. Bureau of Fisheries & Aquatic Resources 34: 4.

Bojos, R.M. Jr., A. Ansula, G. Juanich, R. Arambulo & F.V. Vusse. 1987. The CVRP-I experiment in country-based coastal fishery rehabilitation. Paper presented at National Workshop on Fisheries Policy and Planning, 16–20 March 1987, Baguio City, Philippines. 11 pp.

Castaneda, P.G. & R.I. Miclat. 1981. The municipal coral reef park in the Philippines. Proc. 4th Int. Coral Reef Symp., Manila 1: 183–185.

Cabanban, A.S. & A.T. White. 1981. Marine conservation program using non-formal education at Apo Island, Negros Oriental, Philippines. Proc. 4th Int. Coral Reef Symp., Manila 1: 317–321.

Carolino, V.C. 1987. Under the palm tree. Sea Wind (Ottawa) 1(2): 2–5.

Carpenter, K.E. 1977. Philippine coral reef fisheries. Philipp. J. Fish. 15: 92–125.

Carpenter, K.E. & A.C. Alcala. 1977. Philippine coral reef fisheries resources. Part II. Muro-ami and kayakas reef fisheries, benefit or bane? Philipp. J. Fish. 15: 217–235.

Cruz, R.M. 1986. Philippine coral reefs – a vanishing vital resource. Greenfields 16(3): 35–37.

CVRMP 1984. Nearshore marine fish habitat management. Philippines, Central Visayas Resource Management Project I, Nearshore Fisheries, Annex 10: 1–48 (revised).

Doty, M.S. 1983. Coral reef diversified farming. pp. 437–478. *In*: C.K. Tseung (ed.) Preceedings of the Joint China-U.S. Phycology Symposium, Science Press, Beijing.

Emmerson, D.K. 1980. Rethinking artisanal fisheries development: western concepts, Asian experiences. World Bank Staff Working Paper (Washington, D.C.) 423: 1–97.

Fleras, J. 1984. An overview of the Philippines tropical fish industry. Greenfields 14(12): 50–61.

Fuller, D. 1986. Philippine coral trade update. Traffic U.S.A., World Wildlife Fund 7(1): 6.

Gomez, E.D. 1980. Status report on research and degradation problems of the coral reefs of the East Asian seas. South China Fisheries Development and Coordinating Programme, FAO/UNEP/WG.41/INF.15: 1–48.

Gomez, E.D., A.C. Alcala & A.C. San Diego. 1981. Status of Philippines coral reefs – 1981. Proc. 4th Int. Coral Reef Symp., Manila 1: 275–282.

Gomez, E.D., A.C. Alcala, H.T. Yap, L.C. Alcala & P.M. Alino. 1985. Growth studies of commercially important scleratines. Proc. 5th Int. Coral Reef Congress, Tahiti 6: 199–204.

Goreau, T.F., N.I. Goreau & T.J. Goreau. 1979. Corals and coral reefs. Sci. Amer. 241(2): 124–136.

Hall, H. 1985. Swimming into danger? Int. Wildl. 15(5): 4–11.

Herre, A.W. 1953. Check list of Philippine fishes. Res. Rept. 20: 1–977, Fish and Wild. Serv., U.S. Dept. of the Interior, Washington, D.C.

Heslinga, G.A. & W.K. Fitt. 1987. The domestication of reef-dwelling clams. BioScience 37(5): 332–339.

Librero, A.R., D. Ramos & L. Lapie. 1985. Mechanization: its impact on productivity, cost structure, and profitability of the Philippine municipal fishery. pp. 151–162. *In:* Panayotou (ed.) Small Scale Fisheries in Asia: Socioeconomics Analysis and Policy, International Development Research Centre, 229e, Ottawa.

Luchavez, T., J. Luchavez & A.C. Alcala. 1984. Fish and invertebrate yields of the coral reefs of Selinog Island in the Mindanao Sea and Hulao-hulao in Panay Gulf, Philippines. Silliman Journal 31: 53–71.

McLarney, W.O. 1986. Collecting coral reefs fishes in the Philippines: information and analogue. Scientists debate: more reefs, more fishes die. Annals of Earth (Part 2 of 4) 4(2): 16–20.

McManus, J.W. 1980. Philippine coral exports: the coral drain. ICLARM Newsletter 3(1): 18–20.

Menez, E.G., R.C. Phillips & H.P. Calumpong. 1983. Seagrasses from the Philippines. Smithsonian Contr. Marine Sciences 21: 1–40.

Mines, A.N., I. Smith & D. Pauly. 1986. An overview of the fisheries of San Miguel Bay, Philippines. pp. 385–388. *In:* J.L. Maclean, L.B. Dizon & L.V. Hosillos (ed.) The First Asian Fisheries Forum, Asian Fisheries Society, Manila.

Montalban, H.R. 1927. Pomacentridae of the Philippine Islands. The Philippine Bureau of Science, Monographic Publications on Fishes 24: 1–141.

Murdy, E.O. & C.J. Ferraris. 1980. The contribution of coral reef fisheries to Philippine fisheries production. ICLARM Newsletter 3(1): 21–22.

Nemenzo, F. 1981. Studies on the systematics of scleratine corals in the Philippines. Proc. 4th Int. Coral Reef Symp., Manila 1: 25–32.

Nybakken, J.W. 1982. Tropical communities, symbiotic relationships. pp. 307–382. In: Marine Biology an Ecological Approach, Harper and Row, New York.

Pangilinan, L. 1984. Our imperilled mangroves. Greenfields 14(1): 14–19.

Penaranda, B. 1984. Sodium cyanide: silent killer of the seas. Philippine Panorama, Sunday Magazine of the Bulletin Today 13(43): 32–40.

PNCS 1983. Philippine national conservation strategy. The Haribon Foundation, The International Union for the Conservation of Nature and Natural Resources, The United States

Agency for International Development, The Asia Foundation. 28 pp.

Radan, R.R. 1984. The dismal state of our coral reefs. Greenfields 14(10): 24–32.

Randall, R.A. 1983. Fishery on a fuse. Natural History 92(12): 10, 12, 14, 16.

Robinson, S. 1984. The truth about the use of sodium cyanide in Philippine coral reefs. Greenfields 14(12): 40–48.

Robinson, S. 1987. In defense of the village fisherman: the need for a people-based fishery development policy. Annals of Earth 5(1): 13–15.

Rubec, P.J. 1986. The effects of sodium cyanide on coral reefs and marine fish in the Philippines. pp. 297–302. *In:* J.L. Maclean, L.B. Dizon & L.V. Hosillos (ed.) The First Asian Fish. Forum, Asian Fisheries Society, Manila.

Rubec, P.J. & V.R. Pratt. 1984. Scientific data concerning the effects of cyanide on marine fish. Freshwater and Marine Aquarium 7(5): 4–6, 78–80, 82–86, 90–91.

Sale, P.F. 1980. The ecology of fishes on coral reefs. Oceanogr. Mar. Biol. Ann. Rev. 18: 367–421.

Salvat, B. 1981. Preservation of coral reefs: scientific whim or economic necessity? Past, present and future. Proc. 4th Int. Coral Reef Symp., Manila 1: 225–230.

Silva, P.C., E.G. Menez & R.L. Moe. 1987. Catalogue of Philippine benthic algae. Smithsonian Contr. Marine Sciences 27: 1–179.

Silvestre, G., R. Regalado & D. Pauly. 1986. Status of Philippine demersal stocks – in references from under utilized catch rate data. pp. 47–96. *In:* D. Pauly, J. Saeger & G. Silvestre (ed.) Resources Management and Socio-Economics of Philippine Marine Fisheries, Tech Rep. Dept. Mar. Fish., College of Fisheries, Univ. Phil. Visayas, Diliman, Quezon City.

Silvestre, G.T. & S. Ganaden. 1987. Status of Philippine demersal stocks: an overview. Paper presented at National Workshop on Fisheries Policy and Planning, 16–20 March 1987, Baguio City, Philippines. 26 pp.

Smith, I.R., M.Y. Puzon & C.N. Vidal-Lubunao. 1980. Philippine municipal fisheries: a review of resources, technology and social economics. ICLARM studies and reviews (Manila) 4: 1–87.

Springsteen, F.J. & F.M. Leobrera. 1986. Shells of the Philippines. Carfel Shell Museum, Malate, Metro Manila. 317 pp.

Stoddart, D.R. 1981. Coral reefs: the coming crisis. Proc. 4th Int. Coral Reef Symp., Manila 1: 33–36.

Tangley, L. 1985. A new plan to conserve the earth's biota. BioScience 35: 334–336, 341.

Wells, S.M. 1981. International trade in ornamental corals and shells. Proc. 4th Int. Coral Reef Symp., Manila 1: 323–332.

Wells, S.M. 1982. Coral smugglers, false wealth of the reefs. Oceans 15(6): 55–67.

White, A.T. 1986a. Effects of protective management in coral reefs. Greenfields 16(1): 34–35.

White, A.T. 1986b. Marine reserves: How effective as management strategies for Philippine, Indonesian and Malayan coral reef environments? Ocean Management 10: 130–159.

White, A.T. & G.C. Savina. 1987. Community-based marine reserves. A Philippine first. pp. 2022–2036. *In:* O.T. Magoon, H. Converse, D. Miner, L. Thomas, D. Clark & G. Domurat (ed.) Coastal Zone '87, Amer. Soc. Civil Engineers, New York.

Wilson, E.O. 1985. The biological diversity crisis. BioScience 35: 700–706.

Yao, C.E. 1987. The ecological and economic value of mangroves. Greenfields 17(1): 18–22.

Zich, A. & S. McCurry. 1986. Hope and danger in the Philippines. National Geographic Mag. July: 76–117.

Environmental Biology of Fishes Vol. 23, No. 1–2, pp. 155–157, 1988

European lampreys update

Edward Kott
Department of Biology, Wilfrid Laurier University, Waterloo, Ontario N2L 3C5, Canada

J. Holčík (editor). 1986. The freshwater fishes of Europe, Volume 1, Part I, Petromyzontiformes. AULA – Verlag, Wiesbaden, Federal Republic of Germany. 313 pp. (Can.) $179.00.

This book is the first of a series on the freshwater fishes of Europe. The expressed purpose of this series is to cover all aspects of taxonomy, morphology, ecology, and relationships to man of the species occurring in Europe. It was felt that series such as this are especially needed considering the ecological problems of today and the conservation of species declining in numbers. In order to achieve this goal each species account has been treated as a monograph. All the species of European lamprey have been included in this volume with special attention being given to their classification. The species accounts are the most detailed available and hopefully will serve as models for similar treatments of species from other parts of the world.

The volume is divided into two major parts. The first is a general introduction to the systematics and biology of lampreys. The second is a systematic treatment of the individual genera and species. The general introduction to lampreys is very extensive and completely self-contained. The reader gains a rather complete understanding of lamprey biology, without the necessity of referring to other publications such as Hardisty & Potter (1971).

Section 1 of the introduction covers the morphology of lampreys, both external and those changes which occur as sexual maturity approaches. This is an excellent introduction to the general features of lampreys, however, because of its generalities, the descriptions may on occasion lead to some misconceptions. For example, the lateral line is said to consist of small pigmented pits, but these pits are not pigmented in all species. In the descriptions of changes due to sexual maturity, changes in body proportions are discussed but one of the causes of these changes, a shortening of the body during the spawning migration, is not mentioned. Also, the profound changes that occur in southern hemisphere lamprey are described, especially the development of an enlarged gular sac in males. Although not as striking, a slightly enlarged gular sac also forms in male sea lamprey, *Petromyzon marinus*, during spawning.

The section on morphology is followed by a detailed treatment of identification criteria. As is pointed out, adult criteria are largely quantitative in nature such as counts of teeth, velar tentacles and myomeres whereas in ammocoetes the characters are more qualitative in nature and therefore their identification is more difficult. It is also stressed that morphometrics have to be used with care since they change with size and stage of life cycle. To use measurements one must use individuals of the same size and stage. One inconvenience in this section is the explanatory diagrams. Abbreviations are used on the figures but the meaning of the abbreviations is only given in the 'list of symbols and abbreviations', and not associated with the diagrams.

This is followed by sections on general zoogeography and evolution. The evolution section is useful for indicating our incomplete understanding of lamprey relationships, as both a phylogeny of modern genera and a discussion of the speciation of nonparasitic species (satellite species) is clearly presented. Next there are sections on ecology, ontogeny, feeding and migration which outline most of the important aspects of the life history of lampreys. In these sections important topics such as the relationship of temperature to spawning, and stages in metamorphosis are discussed. Up-to-date information on feeding is also included. The ammocoete is described as a suspension feeder trapping particles by mucus entanglement. This mucus is produced by the gills and pharyngeal walls rather than by the endosytle as once thought. The particles are largely diatoms and desmids, however it is pointed out that the importance of protozoa may be underestimated. Adult feeding is of two types: the large parasitic species are suctorial feeders, feeding on body fluids, whereas the smaller parasitic species may have a biting action and be particulate feeders. *Eudontomyzon danfordi* may even feed on dead fish.

The next section is on a population dynamics and structure discussion sex ratios (males usually predominate in spawning stocks) and mortality, and providing some evidence that changes in population size may be periodic. The introduction ends with a discussion on economic importance. This is useful for, from the human standpoint, this is always of great interest. Also this section emphasizes one of the stated purposes of this series. The great decline in numbers of all species in many of the river systems and the causes are discussed. As pointed out in the next part of the book some species are already protected by law in countries such as Czechoslovakia.

The second part, the bulk of the text, is the section on systematics. Recent concepts have been incorporated at all levels of the taxonomic hierarchy. There is an emphasis on differences between groups which leads to a greater splitting at various levels than one would find if a more traditional approach were taken. An excellent feature is that there are individual keys to each taxonomic level – family, genus, species – rather than an all inclusive key. Each species account is monographic in nature and is discussed under the following topics: synonyms (which are quite complete), holotype, etymology, descriptions, morphology, karyotype, protein specificity, sexual dimorphism, variations, age and size variability, subspecies, hybrids, distribution, introductions, ecology, reproductive biology, important parasites and diseases and economic importance. Topics such as ecology and reproductive biology are further subdivided under headings. The bibliography for each species is generally very complete.

One feature of lampreys usually not discussed in detail which herein is given full discussion is hardiness, one of the subheadings under ecology. Under this heading are discussed not only tolerances of species to various environmental factors such as temperature, salinity and pollutants but also the ability to keep various species in an aquarium.

Although the individual species' treatments are excellent, a few comments concerning some of the species might aid in understanding or prevent slight misunderstanding. In the description of *Caspiomyzon* the number on endolaterals (inner lateral teeth) is given as four on each side. This is the usual statement that is made for the species. However the position of the anteriormost endolateral in *Petromyzon* is anterior to the supraoral lamina. If the enlarged cusp in the same position is considered by analogy to be an endolateral then *Caspiomyzon* has five per side. A valuable contribution has been made in the revision of the genus *Eudontomyzon* with the reduction of many previous species to synonymy. A minor criticism is in the key to adults. The first couplet completely lacks distinctive characters and so might be particularly troubling to a novice. Inclusion of other characters, especially those of dentition such as difference in number of anterials would be an improvement. Under the description of *Lethenteron japonicum,* Heard (1966) is quoted as stating that landlocked individuals in the Naknek River of Alaska reach a size of only 180 mm. Heard was unaware that this sample included two species, one parasitic and the other an undescribed nonparasitic species (Vladykov & Kott 1978a). *Lethenteron kessleri* with 67–72 trunk myomeres is said to agree in this with *L. reissneri* (57–72), however Vladykov & Kott (1978 a, b) consider that *L. reissneri* has only 57–63, whereas *L. matsubarai* which also occurs in Japan has 67–72 myomeres. It appears that here also two species may have been confused in giving a range of 57–72 for *L. reissneri.*

Finally the lingual lamina, although often of considerable taxonomic importance, were in some cases not illustrated or not adequately illustrated. In fact one of the important differences between the subfamilies Lampetrinae and Entospheninae is the nature of the transverse lingual lamina.

The final part of the book consists of three indices; one to the scientific names of the lampreys, one to the scientific name of other organisms, and a third to common names. Such indices are of great importance for a work which stresses the systematics of a group, however a subject index would have made the work more easily used by a nontaxonomist. As a final note, in defining the various genera dentition, especially of parasitic species, has been considered primary. There is a recognition that the nonparasitic species are derived from parasitic species and because of a similarity in habits convergence may occur in characters between nonparasitic species of different genera. A lack of understanding of the nature of nonparasitic species has led some (Bailey 1980) to lump distinct genera by overemphasizing characteristics of nonparasitic species. This has not occurred in the present volume.

References cited

Bailey, R.M. 1980. Comments on the classification and nomenclature of lampreys – an alternative view. Can. J. Fish. Aquat. Sci. 37: 1626–1629.

Hardisty, M.W. & I.C. Potter (ed.). 1971. The biology of lampreys, Vol. 1, Academic Press, London. 423 pp.

Heard, W.R. 1966. Observations on lampreys in the Naknet River system of southwest Alaska. Copeia 1966: 332–339.

Vladykov, V.D. & E. Kott. 1978a. A new nonparasitic species of Holarctic lamprey, genus *Lethenteron,* Creaser and Hubbs, 1922 (Petromyzonidae) from north-western North America with notes on other species of the same genus. Biol. Pap. Univ. Alaska 19: 1–74.

Vladykov, V.D. & E. Kott. 1978b. A new nonparasitic species of *Lethenteron,* Creaser and Hubbs, 1922 (Petromyzonidae) from Hokkaido, Japan. Can. J. Zool. 56: 1792–1800.

Environmental Biology of Fishes Vol. 23, No. 1–2, pp. 159–160, 1988

Epilogue

The last meeting

The week before Vadim Vladykov died three of us (Eugene Balon, Ed Kott and myself) visited him in his nursing home in Ottawa. We had been attending the annual meeting of the Canadian Conference for Fisheries Research and Eugene insisted we should see Vadim who was in failing health. The look of satisfaction on Vadim's face when told about the plans to put together this anthology in honour of his 90th birthday would be the highlight of our visit.

Wednesday, January 8, 1986 was a bright winter day; Ed picked us up in the morning in a rental car and drove to Beacon Hill Lodge on Carling Avenue. Vadim's room on the eighth floor, facing south was flooded with sunlight; he was lying on a hospital bed along a wall just inside the door. He looked weak and frail. He recognized Eugene and Ed extending his hands to both and acknowledged my presence by enquiring whether I had ever had the pleasure of eating sea lamprey – he had remembered an article I had written on this subject some twenty years ago. While my companions carried out a halting conversation with their old

This was the view directly in front of Vadim's eyes in his last home.

colleague and friend, I gazed about the room observing some of the books, mementos and photographs he had accumulated over a long, productive career.

Opposite the bed was a wall lined with floor-to-ceiling book shelves containing volumes mostly on fish. A radio and some souvenirs filled the unoccupied spaces. Paintings and plaques, again some with an ichthyological theme were interspersed along the walls with a placé of honour filled with a photographic portrait of Vadim's father.

After a while Vadim was visibly tiring from the conversation and the appearance of a nurse brought an end to our stay. The last goodbye was said, we left room 809 and made our way to the parking lot below.

The sad event of the following week, of course, affirmed that our visit was none too soon. We believe that Vadim died with the knowledge that this friends were showing their appreciation of his accomplishments in the creation of this volume.

Robert W. McCauley

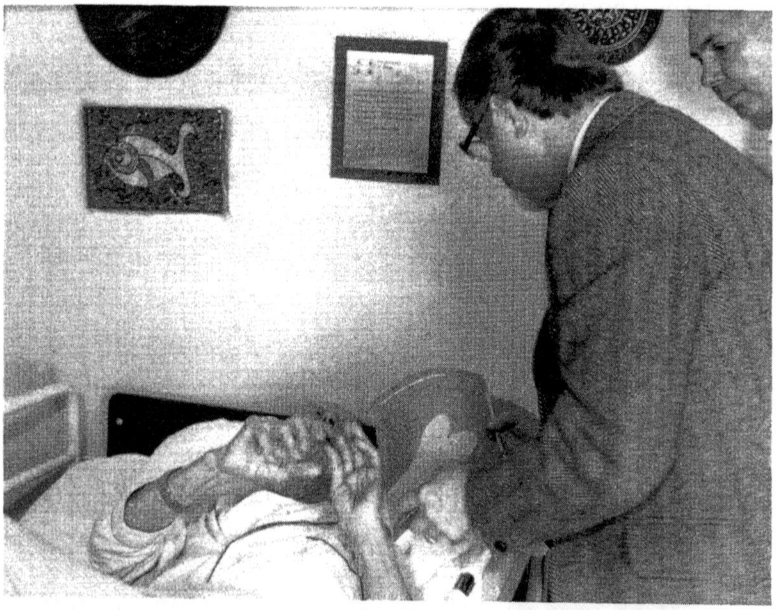

Vadim is told about this volume by Eugene and Ed. Photographs by R.W. McCauley.

Species and subject index

www.ingramcontent.com/pod-product-compliance
Lightning Source LLC
Chambersburg PA
CBHW080934240926
51849CB00003BA/455
9789061936619